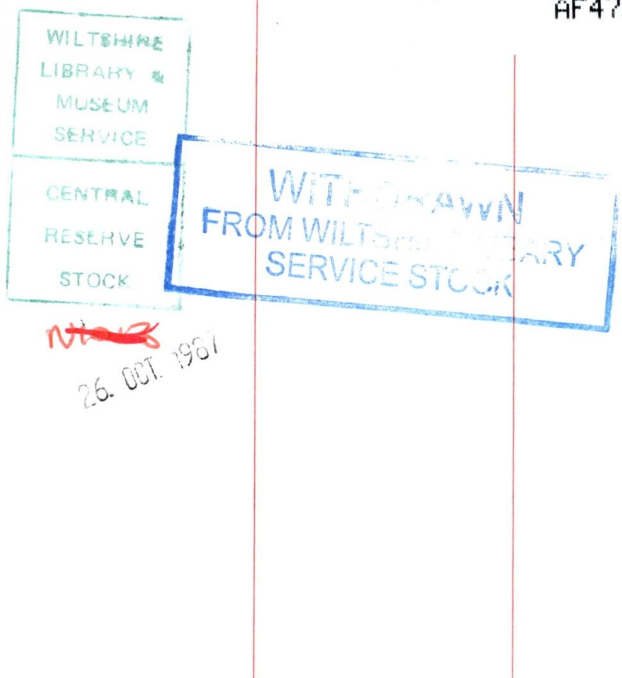
WILTSHIRE
LIBRARY &
MUSEUM
SERVICE
CENTRAL
RESERVE
STOCK
WITHDRAWN
FROM WILTSHIRE LIBRARY
SERVICE STOCK
26. OCT. 1987

0 321832 002

FROM FOREST TO FARM

From Forest to Farm

BRISCOE MOORE

PELHAM BOOKS

First published in Great Britain by
PELHAM BOOKS LTD
26 Bloomsbury Street
London, W.C.1
1969

7207 0134 X

Printed in Great Britain by
Northumberland Press Limited
Gateshead

TO MY WIFE
who shared so much of it

CONTENTS

ILLUSTRATIONS

Pictures Nos. 4, 14, 17, 23, 25, and 30 are reproduced by courtesy of New Zealand National Publicity; Nos. 5, 8, 9, 10, and 11 by the Auckland Institute and Museum; Nos. 16 and 18 by James Aviation; Nos. 24 and 29 by Northland Photography.

PREFACE

Many in Britain will be familiar with the term 'New Zealand Lamb' and 'the sun-drenched pastures of New Zealand' used in the sales promotion of our produce. This is a story of the realities behind these slogans—of one who has created some of these pastures and bred and reared some of the lambs which end up on the dinner tables of the U.K.

Interrupted by World War I, my life as a sheep farmer got into its full stride only in the post war period, with the acquisition of a block of land, mostly in standing forest, in the northernmost part of New Zealand.

I have tried to paint for the reader something of the drama, trials, exultations and disappointments involved in creating a pastoral farm out of the standing bush. To do this the more effectively I have devoted chapters to each significant operation rather than follow a strict chronological narrative.

I have described our New Zealand ways and what a newcomer can expect, and have made some philosophic reflections that may perhaps help some younger readers to chart their course in an uncertain world.

I gratefully acknowledge my debt to all those who helped me through those years, above all to my wife, and latterly my son. My thanks go to David Grace for his constructive criticism and suggestions.

A.B.M.

Chapter 1

LEARNING THE GAME

I should have been an engineer or a shipping man, but a fortuitous two months' holiday on a sheepfarm at the age of sixteen changed all such ideas. The open air life, with a horse to ride and stock to look after, called to some deep instinct within me which set the course of my life from that point on.

My father, a shipping manager, had been sent to New Zealand in 1889 to inaugurate the running of a line of insulated steamships for carriage of frozen meat to Britain. My mother was the daughter of a pioneer in another field who had arrived by sailing ship in the 1860's to extend the trade of an internationally known hardware firm. I was born in Dunedin in southern New Zealand in 1891. This being my family background, I was fortunate enough to have been given a good education, and when I persisted in my idea of farming I was sent to Lincoln Agricultural College in Canterbury, in the South Island of New Zealand, at that time the only Agricultural College in New Zealand.

In those days, the early 1900's, agricultural education was a comparatively new idea—the general farming view was that a young fellow should learn his business the hard way, on the farm, and this newfangled idea that books and learning could help him was scoffed at by many. Nevertheless I buckled down to learn, and the three years 1908 to 1910 spent at Lincoln proved subsequently to be one of the best investments of effort I ever made. We all had a duty, such as milking or grooming a working horse before breakfast, then we did a half-day's work on the farm and another half day attending lectures, about two hours 'swat' in the evening rounded off our day. We had our noses well and truly rubbed into the practical side of farming.

Little did I realise, when I finished my course in 1910, that nearly half of my friends and contemporaries were soon to be swallowed up and killed in World War I. We were all the right age and types to be drawn into the maelstrom which was brewing even then.

Although the main theme of this book is the conversion of a block of forest land in the extreme north of New Zealand, into a farm, I must ask something of the reader's indulgence while I outline the events that went before. So much of what happened to me was typical of the experience of thousands of young New Zealanders.

Upon, as I thought, completing my agricultural education, I realised that I had a lot more to learn on the practical side before I would be capable of taking on a farm of my own. I managed to get a job as a 'cadet' or apprentice shepherd on one of the large back country sheep 'stations' as they were called. This was situated in South Canterbury, in the South Island, and consisted of about 150,000 acres of natural tussock land pasture, upon which, when stocked lightly, sheep did well. Included in the property were thousands of acres of rolling downs and mountain range country running up to 6,000 feet above sea level.

I 'made my marble good' as the saying is, with the head shepherd, to the point where I was taken on as a full-blown shepherd. I was provided with a horse, but had to buy my own team of dogs. My ambition was to own my own horse saddle and bridle as well, so I saved hard, and ultimately the proud day came when I mounted my first young horse, not too certain as to whether she would let me ride her or buck me off.

This large property, nearly all Crown leasehold, was typical of many of the sheep stations that were settled in the early days in the South Island of New Zealand, some owned by companies, many by individuals. Such places would have a permanent staff of say six to twelve men, according to size, who would be accommodated in a bunk-house, while a cook would be employed to run the cook-house. In the summer season extra men, 'musterers' would be taken on to help get in the sheep for shearing, dipping and other purposes. With 50,000 sheep to handle over large areas these men were an essential part of the set-up. They arrived carrying their 'swags' on their own horses, and usually would have four or five dogs each. Some mountain country needed up to twelve men or more to get the sheep in.

On one range we would ride out the night before, with pack mules carrying tents and the cook's gear, and camp for the night. Up at the unholy hour of 2.30 next morning, we would breakfast and set off to climb several thousand feet to our 'beat'. With one man on top and up to six down each side we worked our way along the faces picking up all the sheep as we went. We would sometimes be spaced a good half mile apart. Today the men who do this are able to communicate with walkie-talkie sets. While we shepherds would be mustering, the cook conducted the team of three mules along the bottom beat to the camp site for the night. These were temperamental animals who, surefooted as goats, needed understanding. One had to know for instance that old Jack would not stand for any clanking noise in his load, or too much weight. If this idiosyncrasy of his was not taken into account, he would wait with an angry twitching of his long ears until his load was completed, and then, making a circular track, he would set to and methodically buck off his whole load, splattering pots and pans around in a circle.

Towards the end of the day we would have long strings of sheep moving ahead of us down to the bottom beat. Here they coalesced into a mob of several thousand which we put through a fence dividing off the area we had just cleared. These would be picked up with others, the following day. This mobbing up was a great sight as thousands of sheep milled round throwing up a cloud of dust golden in the rays of the setting sun. As the sheep went through the gate the last act was to grab a good conditioned wether. This was slaughtered on the spot and the meat used for chops and stew for the meal that evening and the following breakfast. The theory was that the meat transported itself to the point where it was needed, though the method left something to be desired, from the point of view of toughness.

All this mountain mustering had to be done on foot. One needed to be fit, and we all carried stout manuka sticks 5 feet long to steady ourselves as we scrambled across shingle slides and rocky outcrops. It was very rough walking and the dogs used to develop sore feet from the abrasive effect of the rock on the pads of their feet. None the less it had an immense appeal to a young man. To reach your beat 4,000, 5,000 or 6,000 feet up in the cool clear air of early morning and see the rays of the rising sun tinting all the tops of the mountains around you with a rosy glow, while the valleys remained in deep shadow, was something never to be forgotten. There were too, the mountain parrots, the keas with their green plumage and flame red breasts and wing undersides. They are very intelligent birds and became a pest because of their habit of alighting on a sheep's back and pecking their way through the wool to reach the kidney fat, thus killing the sheep. They have also been known to act in unison running sheep together in a small mob and then forcing them over a precipice.

On one occasion I had the top beat, and had to wait for the men below me. I sat down and presently had seven of these birds around me in a circle. As I kept quite still they hopped in closer and closer and we had a most amusing session of interviewing. The second day's muster was a repetition of the first, with more and more sheep moving before us. Then, usually in the early afternoon, at the point where the mountain range terminated, the 'leads' from each side came together and were gradually mobbed up until we had perhaps 12,000 sheep in one large mob. These were marshalled into a long narrow column, and with one man behind and several along each side we set off for the yards, some miles away, that were our destination.

There was much skill and hard work involved in all this and one's team of dogs was fully extended. There were 'do's and don'ts' and conventions that the young shepherd had to learn—also dangers. On one occasion, as two groups of several thousand sheep met at the end of the range, a noisy young dog got out of control, and with his barking caused two lots of sheep to stampede together at a point where there was a small depression. As they crowded into this in panic, others piled on top of them, and in a matter of a minute or so about 200 sheep were killed by smothering. This kind of thing could happen quite easily with the Merino mountain sheep which were very scary.

A source of trouble were the wild hermit sheep. These had dodged successive musters, often had two or three years' growth of wool, and because of this were nearly wool blind. They would lead other sheep off in the wrong direction, usually ending up in some inaccessible spot where to put a dog in after them was to risk putting them over a small precipice. Occasionally we got one in but they usually had to be shot.

Chapter 2

LIFE ON A LARGE SHEEP STATION

These back country stations were a repository for all sorts and conditions of men. The active stock work was carried out by men mostly in their twenties and thirties, there to gain experience and earn money. They were known as 'the lizards' (because they were rock-crawlers I suppose) and were usually unpopular with the cook because they required meals at all sorts of odd hours according to their work. Many of the other jobs were filled by older men of a quite different calibre who had gravitated there from all over the place and who liked the back-country life. Some of these were 'remittance men' often of good family whose presence was not desired at home. Many had adventured all over the world as sailors, soldiers, miners and what have you. These filled the posts of carpenters, blacksmiths, stablemen, gardeners, fencers, and so on. There were many educated men among them.

I recall one who was the station carpenter and handy man. He had been amongst other things a boss in the De Beers diamond mines in South Africa. His trouble was drink. He would periodically have a bad bout, when he would verge on delirium tremens, and I would be called in to pacify him and assure him that the snakes and reptiles threatening him from the wall were not really there.

A colourful character on the station was 'Buffalo Bill' the bullocky. He operated one of the two bullock-waggons which carted most of the station's baled wool to the rail-head on the first stage of its long journey to London for sale. Travelling at a slow but steady two miles an hour, the waggons carried thirty-six bales each weighing about 3 hundredweight. At the end of the day's journey making camp was a simple matter of arranging a tarpaulin from the floor of the waggon to the ground. Under this he slept while his bullocks grazed either on the road-side or in an adjacent paddock. His pony, which led behind the waggon during the day, was hobbled at night and used in the morning to round up the bullocks.

This pony, a wall-eyed chestnut, was just about as hard a case as his master, and it was not uncommon for him to put on a buck-jumping display on being mounted in the morning—this was a free show for the station hands when it took place in the paddock in front of the cook-house at the homestead.

Once at Christmas time he had been entertained by some friends, and he arrived back at the station rather the worse for wear and without his dentures. 'What happened to your teeth, Buff?' asked one of the men. 'By golly,' said Buff, 'I must've left them on Mrs. 'Obb's mantelpiece.'

Shearing, removing the 'golden fleece'—now alas not so golden, was one of the great occasions. This station had two woolsheds, each equipped with twenty machine stands. The largest could cover 1,600 sheep overnight. They were spaced 20 miles apart to avoid having to drag the sheep excessive distances to have their wool removed. We shore all the sheep at the southern end of the place at the homestead shed, and then moved everyone up to the other shed. Shearers were recruited from wide areas, many of them being Australians attracted by the long 'run' and consequent cheque which could be earned when the shearing of 50,000 sheep was divided between twenty men. It was a great sight to see the shearing shed at full pitch. From twenty stands, ten down each side of the shed, the creamy fleeces were being peeled off the sheep. Shed hands, a woolclasser and pressers in turn dealt with this fascinating harvest until it ended up in its appropriate lot in a bale branded for sale in far off London.

We shepherds and musterers were kept going flat out bringing in the sheep from distant blocks and taking them back to their pastures as each lot was shorn. If the weather was hot it was hard on sheep and men, and it was a wearing drawn-out business forcing along, single-handed, three or four hundred bawling woolly ewes with their lambs, over open tussock country perhaps four or five miles to the woolshed. It was always much easier taking them out after shearing, as the ewes, relieved of their wool, were keen to get back to their own haunts. Hill sheep, particularly the merino, have a very strong sense and instinct for what they regard as their own country.

Another busy time was dipping, when every single sheep, most of whom objected strongly, had to be heaved into a bath containing an arsenic sulphur mixture as a protection against the ticks and lice which would otherwise infest their fleeces. During the lambing season each of the permanent shepherds was made responsible for a 'lambing round' which might cover 12–1,500 ewes grazing on perhap 2,000 acres of land. Day after day, Sundays included, he performed the offices of a midwife to the ewes under his care throughout the two-months' period of lambing. I shall enlarge further on this work in a subsequent chapter.

In January and February came the time for mustering in and drafting out the surplus sheep to be sold, which, with the wool, provided the gross revenue from the property. These included about 5,000 fat lambs for the London market. These would be picked out and marked by the buyer. I remember feeling full of importance on the one and only time when I was sent to order a special train to take 1,500 lambs to the works.

In the autumn, in late April or May, would come the 'Fall Muster'. Incidentally I never found out how this American term for autumn came to be used in the back country of New Zealand. All these stations in the mountain tussock lands would be divided by fencing into what was known as 'summer' and 'winter' country. The summer country was mostly comprised of the higher ranges and those lying away from the sun where the winter snowfall was heavy. Sheep were run on this range country through the summer months only, and brought down to the lower country—the winter country—in the late autumn. This consisted of rolling downs, the lower hills, and faces lying to the sun where snow did not lie. Even so, exceptional snows every few years could, and still do, catch the high country man out and cause heavy losses of stock. This snow risk is an accepted hazard of high country farming in New Zealand and such losses are a heart-breaking business for those concerned.

Many of the larger stations that I knew have since been cut up into smaller properties, but these mountain lands do not lend themselves to close subdivision and must have a generous acreage to be an economic unit, and survive the 'snow risk' and other hazards. Their carrying capacity is low, often one sheep to four or five acres or more, and although some improvement work is being done on the better types of tussock land, many will remain as straight out mountain grazing on which little improvement is possible. Having given a brief description of the main activities of the year I could add that although most of our work was with the stock, we did have interludes of horsebreaking, cutting chaff for the working horses, repairing fences, or anything else that had to be done.

Chapter 3

I BECOME A 'RUNHOLDER'

After three years of this I thought I had achieved a fair competence in all this kind of work, and was rash enough to contemplate launching out on my own. I was twenty-three, fit, full of energy and with a flattering concept of my capacity. The best method of starting as a 'runholder' was to acquire what was known as a 'small grazing run'. These were places of 1,500 acres upwards, often subdivisions of larger properties. They were Crown leases of twenty-one years with the right of renewal at modest rentals, and when first put on the market were so much sought after that they had to be balloted for. I tried this balloting, but when there were as many as eighty applications for one run, concluded it was pretty hopeless. I therefore made up my mind to buy a place. This meant paying the owner for the stock, the improvements on the place, and the 'goodwill' or value of the unexpired period of the lease.

By paying over all the cash and savings I could scrape up, plus some family assistance, and borrowing heavily on mortgage, I was able to finance myself into a small grazing run in the district in which I had worked. This was a block of fairly steep tussock hill country, the northern side of a small range of hills running up to 4,000 feet. It was four miles long and a mile wide, 2,400 acres in extent, and was known as 'Ben Lomond'. It had a house and outbuildings, woolshed and sheep yards and some miles of fencing. At the time I bought it as a 'going concern' there were 1,600 sheep on it. I established myself with a cadet, a young fellow anxious to learn, with a housekeeper to look after us. This all seemed to me a great event at the time, but I was soon to find that my judgement of the property was not that of the knowledgeable sheepfarmer I had imagined myself to be.

The joy of ownership was soon to be tempered by a realisation that the tussock pastures were badly eaten out, and that in fact the place was overstocked with the 1,600 sheep then running on it. I ultimately concluded that 1,300 was the maximum it could safely carry, and reduced the stocking to that figure. In going through the sheep to do this I had a further disappointment when I found their quality was below that of my original estimation. All this of course was traceable to inexperience, as were other things that were to give me concern. A more experienced man for instance would have detected the signs of overstocking that I, in my eagerness to acquire a place, had overlooked. These were dearly learnt lessons—I am glad to say that in later years they helped me prevent other young men making similar bloomers when they asked my advice.

I resolved to make the best of a bad bargain and upgrade the place by careful breeding to improve the stock and woolclip. In this I was ultimately successful, but found that this property did not offer enough scope for my ideas and energy. It already had all the buildings, yards and fences that were needed, and the possibility of any serious improvement of the native tussock pasture that had been permanently damaged by over grazing was nil.

My first lambing percentage was poor in the extreme, some 30 per cent, due to the indifferent management of my predecessor. Then I struck one of the worst dry seasons for years, and in another year when I had expectations of a good lambing lost some 400 newborn lambs in a period of heavy cold rain. I was learning, the hard way, that despite the best care and attention, nature, in the form of fickle weather, can deal out some heavy blows. The inference to be drawn is that in any farm budget or estimate of returns the item 'contingencies' should figure prominently. In all I was to work 'Ben Lomond' for three years. I had my ups and downs, but had no trouble with the day to day work for which my previous years had fitted me.

Throughout this time, the threat of a world war was looming. In 1911 the New Zealand Government had brought in compulsory military training, and all we young men were roped in and organised into units of a newly created territorial army. I found myself overnight a trooper in the Otago Hussars, a regiment of Mounted Riflemen recruited from the country areas. We had to provide our own horse, saddle, and gear, but were issued with uniforms and rifles. We attended two camps each year, one of a fortnight in the autumn, and another of a week in the spring. They were timed to cause the least possible interference with the seasonal farm work.

At first there was a good deal of vocal opposition to compulsion, and some were sent to gaol for defying the law. But the majority of us recognised the ominous portent of events overseas and were glad to do what was asked of us. Gradually an *esprit de corps* was built up in the different units, and we all became keen to turn ourselves out well and do our regiment credit. We took especial pride in our horses and saddlery, most of the men being from farms, stations, and sheep runs, over a large area. The touch of discipline and the fact that we were standing up to be counted as defenders of our beautiful land had an excellent effect upon the morale of our youth. It also made possible the early contribution of trained and equipped forces that New Zealand made upon the outbreak of World War I.

The outbreak of war in August 1914 found me tied to a chair with one leg in plaster of Paris as the result of an accident on the farm. Most of my contemporaries were early volunteers, and many were soon to perish or be wounded in the ill-starred Gallipoli campaign. These grim undertones to come were not appreciated by the crowds of young men, many of them idealists, who thronged the recruiting offices. I felt badly out of it for I had lived, worked, and trained as a soldier with them, and had to see them go off to high adventure while I remained behind as a temporary cripple. I consoled myself with the thought that I would soon catch up with them, but the results of my accident left me with a permanent weakness in one knee, and the doctors refused to pass me for active service.

I returned to 'Ben Lomond' and carried on single-handed. My assistant, a splendid young fellow, had volunteered, and was soon to be killed. Being something of an idealist, this watching of so many of my friends lose their lives in defence of what they believed in while I could make no contribution, had a depressing psychological effect on me. I tried again and again to be accepted for overseas service, but it was not until 1916, by which time there was some relaxation in the standard of fitness, that I finally persuaded a somewhat doubtful medico to O.K. my papers. I had enough time before reporting in camp to put in hand the disposal of my horse, put the property up for sale, and arrange for the best two of my dogs to be cared for while I was overseas.

I was fortunate in being able to sell 'Ben Lomond' as a going concern at a price that allowed me to recover my original capital. This sale, to a man only a few years older than myself, who had been rejected for service, had an ironic sequel. I went off to war, serving the next three years as a soldier during which I had some narrow escapes and four months in hospital, but returned to New

Zealand in one piece. My successor on the property got caught in one of the heavy snows that every few years devastate the high country, and while attempting to rescue some sheep in a freezing blizzard, got himself frozen to death before he could be rescued. One would have expected the odds against survival to have been more heavily stacked against me rather than him.

Chapter 4

MOUNTED RIFLEMAN IN PALESTINE— SOME NEAR MISSES

Since the start of compulsory military training in the years before the war I had risen through the various non-commissioned ranks to the point where I had sat my examination for a commission. In this I had been successful. So it was that I went into camp as a young officer in charge of a group of N.C.O.'s. We were the nucleus of a squadron of 130 men due to arrive in camp when we had concluded our intensive training. It was to be our job to knock them into shape with the assistance of a few staff sergeant-majors from famous British cavalry regiments.

This was an intensely interesting experience. I had some aptitude for it, worked long hours, got to know all my men individually, and ultimately won their confidence. Although predominantly of rural origin, they came from every walk of life. There were farmers and station workers, shepherds, teamsters, blacksmiths, barmen, policemen, warders, business men and what have you. Most of them had had something to do with horses, and could ride, but a few could not cope, and had to be weeded out in riding tests. These tests also offered the opportunity, which I used without scruple, to weed out the odd one or two 'bad hats'.

The sudden transition from civil to military life caused difficulties for many, who came to me with their problems. I found myself involved in every role from financial adviser to marriage counsellor. I flatter myself that I saved at least one marriage from going on the rocks.

It was a fascinating business gradually to weld all this human material into an effective military unit fit to reinforce the New Zealand troops overseas that had made such a name for themselves. After four months' training we were ready to go overseas and were embarked at short notice at midnight on a stormy wet night in the steamer *Monowai* en route to Sydney. There we transhipped and joined some Australian troops, making 900 of us in all on the troopship *Port Lincoln.* We set sail for Egypt via Colombo. Approaching this port we had an outbreak of cerebro-spinal meningitis on the ship, and we were hurriedly put into camp ashore to allow all positive cases of infection to be detected and sorted out.

This gave us an entrancing ten days in Colombo. We were particularly fascinated by their picturesque little hooded carts pulled by two little bullocks yoked to the pole which seemed to be a chief means of transport. Ricksha rides (including some unofficial races) were popular, as were bananas, pawpaws, and other fruit at very low prices.

When those destined to stay behind for further observation had been sorted out, there arose the question of which one of the seven New Zealand officers was to remain behind in charge. Amongst us only the eldest man was married, with a family, and we others tried to arrange that he stay. He would have none of this, and insisted that the matter should be settled by drawing lots. This was done, and whether by design or accident, he was left with the shortest straw. The sequel to this was tragic. Although he arrived in Egypt a long time after us, his arrival coincided with a heavy casualty rate. He was sent straight up the line as a replacement, and was killed almost immediately.

We ultimately reached Suez and entrained for Ismailia on the Suez Canal. Nearby was Moascar camp, the base training depot for the New Zealand Mounted Rifle Brigade, which formed, with two Australian Light Horse Brigades the Anzac Mounted Division. This division, at the time we arrived was in front of the enemy line at Gaza. The strength of our New Zealand force was 1,850 men and 2,200 horses. Our New Zealand Brigade was essentially a front line fighting force in which there was a steady wastage both from battle casualties and tropical disease. The latter took heavy toll in the later stages of the campaign in the Jordan Valley, where malaria was endemic. The Brigade was maintained at full strength throughout the war by a steady stream of trained reinforcements from New Zealand.

To maintain 1,850 men in the field for the full period of the war over 10,000 men had to give their services to our New Zealand Mounted Rifle Brigade. A striking commentary on the wastage of war. We formed part of the Egyptian Expeditionary Force, composed of over 400,000 British and Commonwealth troops. All the mounted troops together formed the famous Desert Mounted

Corps, said to be one of the largest forces of horsemen ever controlled in one command, some 50,000 in all. This is not the place to give a detailed account of a campaign that has been recorded elsewhere, though I propose to mention a few highlights as they affected me personally. I was late on the scene, and cannot claim the long experience of front line service that many there before me had. None the less I had enough of being shelled, machine-gunned and shot at to last me for the rest of my life.

The worst fight I was in was the first attack on Amman, now the Jordanian capital, in which the British forces were beaten off with heavy losses. From ground nearby we saw a classic attack by the ill-fated Camel Corps wither away under concentrated machine-gun fire. We were heavily involved ourselves in an attack on the same Turkish position soon after. When one bullet went through my greatcoat, another smashed the prismatic compass I carried, and a third pierced my haversack in the one day, I thought the wings of the Angel of Death were brushing too close for comfort. There was always some macabre humour, as in the case of one man who had a bullet through his felt hat (we had no steel helmets) and the tip of the lobe of one ear shot off. He was most indignant because the M.O. refused to evacuate him as wounded!

Another incident that had its element of grim humour was during one stage of our attack on the Turkish position. We were lining a rocky ridge which gave us some cover from the heavy shelling of about fifteen enemy guns trying to dislodge us. In our rear was about 400 yards of open country with, beyond it, a second ridge of rocks amongst which were hundreds of arabs watching the free show. Without warning, the enemy gunners, who were often Austrians, and very expert, lengthened their range, and plunked twenty or thirty shells amidst the Arabs behind us. We were able to observe, with some satisfaction, Arabs scuttling at high speed in every direction.

The most hair-raising stunt I took part in was when volunteers were called for to cut the Turkish railway line north of Amman. This was to prevent the retreat northwards of thousands of enemy troops concentrated there in the concluding stages of the campaign. As a night operation it involved finding our way in the dark across fifteen miles of rough unknown country, and evading the enemy outposts spaced along the railway to protect it. This we successfully accomplished, breaking the line by uncoupling the fishplates at a point where it was unlikely to be noticed, and without detection by the Turks. On the return journey, leading the advance guard with my oil bath prismatic compass as our sole guide for direction, I suddenly found we were off our course and far too close to the enemy lines. It was bitterly cold, two in the morning, and just in time I woke up to the fact that the low temperature was congealing the oil in the compass and inducing a dangerous error in the bearings it recorded.

When we captured Amman with a large haul of prisoners a day or two later, we sent a patrol up the railway and there, sure enough, was a first class train wreck strewn along the embankment where we had cut the line. Since then I have always regarded myself as something of a connoisseur on how to wreck a train, but have had no further calls on my services in this connection. This of course was on the famous Hedjaz railway on which Lawrence had a good deal of experience with demolitions. On one occasion, when he detonated a heavy charge electrically from a distance away, he recounted that one of the heavy driving wheels of the locomotive landed with a loud clang on the rocks behind him!

As the end of the war approached the Turks were steadily driven back and large formations were cut off by the use of fast moving units such as ours. Following our capture of Amman, a main centre and engine depot on the Hedjaz railway, our divisional general received an extraordinary message from the Turkish general in command of about five thousand enemy troops at a station called Ziza fifteen miles below Amman. This was to the effect that he and his troops were prepared to surrender to the Anzacs if and when we could afford them protection from hordes of armed Bedouin Arabs who had surrounded them. We New Zealanders were sent off on this strange assignment. Within a few hours, pushing our horses along in the dust, we had reached Ziza, and very quickly threw an armed ring around it. We had had about enough of the Arabs with their thieving cut-throat tendencies, and took pleasure in hastening their departure with a few shots placed where they would do the most good. We had found by experience that their speciality was looting—when somebody else had done the fighting.

In the railway yard at Ziza were three complete trains, masses of military equipment including guns, with dead and dying men here and there amidst an appalling stench resulting from the insanitary habits of the Turkish troops. Their sole source of water was a nearly empty concrete cistern with a corpse lying half in, half out, of the water.

This was the finale as far as we were concerned, shortly to be followed by the armistice. The war was at an end. In its later stages we had spent much time in the malarious Jordan Valley, before the final capture of Amman in the high hills to the east of it, the land of Moab. Apart from battle casualties, we had had heavy losses from malaria, many men dying from the malignant type of this loathsome disease. It was tragic, on the point of victory, to see so many, who had survived so much of the fighting, lose their lives like this. Wounded and sick were evacuated through a succession of casualty clearing stations by hospital train on the desert railway that had been built behind the front as the troops pushed the enemy northwards through Sinai and then Palestine. In the last stages Jerusalem was the rail-head, but it was a terrible journey by road ambulance from the Mountains of Gilead down into the Jordan Valley and then up the winding mountain road from Jericho, 820 feet below sea level, to Jerusalem 2,500 feet up in the hilltops. The ambulances of those days were not the comfortably sprung vehicles fitted with shock absorbers that we have today, and such a journey was an excruciatingly painful experience for a badly wounded man.

At one stage I collapsed with an undiagnosed disease, and was carted off on a stretcher with a large label marked P.U.O. tied on to me. This meant 'Pyrexia (high temperature) unknown origin'. As I lay in the casualty clearing station (tended by nuns) at Jerusalem, and watched the other malaria patients having large doses of quinine injected through their sides into the spleen, I wondered what lay in store for me. I remained a medical mystery until I got to the hospital at Gaza, where a somewhat alert M.O. asked: 'Well, what's the matter with you?' I was getting a bit fed up by then to put it mildly, and replied rather tartly and rudely: 'That's what you quacks don't seem to know.' He peered closely into my eyes and then said: 'You're going to be a fine case of tropical jaundice—the whites of your eyes are already a canary yellow.'

So it was—I was moved by hospital train down to Kantara on the Suez Canal, and then on to base hospital in Cairo, where two months of treatment put me back on my rather shaky legs. I remember with gratitude the devoted attention of the nurses in the clearing stations, the hospital trains and hospital.

Chapter 5
ARMISTICE

After the final enemy collapse we made our slow way back in stages to Ayun Kara, a Jewish settlement eight miles inland from Jaffa. Our travel had to be slow because our ranks had been so thinned by malaria that we had barely enough men left to handle all our horses and gear. In my own troop of thirty-six men and forty horses I had left only a dozen men, so that each man riding his own horse had to lead two or three others. Of the six officers, only the O.C. and myself were left.

Ayun Kara, which had originally been captured by the New Zealanders after a heavy fight with a superior Turkish force, when we had had many casualties, was a familiar spot to us. We were regarded by its Jewish colony as their liberators. There were hundreds of acres of vineyards, centred upon a large wine-press financed by the Rothschilds. There were large stocks of wine held to mature in huge hogsheads and glass-lined tanks. This was freely available at very reasonable prices and it probably did us all a lot of good. But we had to see that things did not get out of hand. Accordingly a twelve-man twenty-four-hour guard was mounted on the wine-press. It was understood that the guard, in their period of duty, was to be allowed a liberal ration of free wine. This was, in my army experience, the only really popular guard duty I had noticed. The changing of the guard daily was a ceremony worth watching. Lined up, the old guard looked slightly shopsoiled and not too steady. The new guard facing them were smart, with lots of spit and polish, and literally licking their lips with anticipatory relish!

Later we moved to another camp, at Rafa on the southern border of Palestine, there to await embarkation for home. I wonder to what extent the now intensely nationalistic Israelis realise that their country was liberated from 700 years of alien rule solely by British troops, with a high cost in lives. Throughout the long campaign, starting at the Suez Canal and ending at Damascus, the

Australian and New Zealand mounted troops were always in the forefront of the fighting, and took rather more than their share of it. As exemplifying this, the official casualty returns for three mounted rifle regiments, Auckland, Wellington, and Canterbury totalled 3,254. Of this number, 1,100 from our small distant country lost their lives.

Now the fighting was over, all eyes were turned towards home, and never did Aoteroa, New Zealand, the 'land of the long white cloud', appear to be so desirable. Our careers, whatever they were, had been interrupted by our period of service, and we wanted to get out of the army and re-establish ourselves in civilian life.

In this we had to wait our turn—shipping was in short supply and armies had to be repatriated from all over the world. In the interim we set up an educational course in a number of subjects designed to help the troops on their re-entry into life at home. This was quite popular and served to mitigate the boredom of the men throughout several months. In a weak moment I had undertaken to lecture on sheep-farming, and was somewhat rocked, to find myself facing an initial audience of four hundred men, to whom I had to talk daily. It was quite a challenge—knowing there were many experienced men in the crowd I kept to fairly sound basic stuff, and then we had a lot of fun in a kind of free for all discussion group in which I encouraged them to propound their ideas with myself as chairman!

One harrowing experience in the months following the armistice was the disposal of 800 of our older horses which were nearing the end of their useful lives. They had carried us faithfully and well, and were our good friends for whom we did not hesitate to 'pinch' extra food when we could get it. They could not be returned to New Zealand and we could not bear the thought of leaving them in their old age to the cruelties of the Arabs or Egyptians. It was accordingly decided to shoot them, and daily parties were detailed to lead them out to the coastal sandhills and do this dreadful job.

Our stay in camp at Rafa was rudely interrupted by the outbreak of rebellion in Egypt, with a rash of train-wrecking and burning of public buildings. We were hurriedly entrained for the Suez Canal, where we were re-equipped with fresh horses and gear, and then sent on to the Nile Delta. En route we passed the wreck of the train that had preceded us. This was in the nature of a pacification job, a show of force in the main towns of the delta, in which the malcontents were rounded up and thrown into durance vile. It

The author's troop of mounted riflemen in World War I.

Captured Turkish cavalry officers being escorted to the rear by New Zealanders.

New Zealanders crossing the River Jordan at the site where later was built the Allenby Bridge.

Felling a forest tree in the bush.

A typical bushman's 'whare' built of split palings.

was interesting to those of farming background to get such a good look at this area of rich alluvial land irrigated by water channelled from the Nile. Our horses too loved it, and we had a problem checking the men from over-feeding their neddies with the rich berseem clover the fellahin grew for their own stock, which were all tethered.

The rising was suppressed quietly but firmly, and military courts were set up to try the troublemakers. We were by then encamped near a fair-sized town complete with a court-house and gaol. Another New Zealand officer, myself and one Egyptian police officer were appointed as the judges to try the local offenders, forty in all. The evidence in every case was conflicting, to put it mildly, and we quickly realised that many of the witnesses had been bought and were monumental liars.

One rather fine-looking man of good physique had a whole string of crimes listed on his charge sheet ranging from wrecking a train to burning a government building. The chief witness against him was the chief of police, who tried to damn him in no uncertain terms. Other witnesses, of course, protested his innocence. During his trial we had an adjournment, and the police officer who was a member of the court then undertook to enlighten us. He apparently knew the whole story, the gist of which was that the rather good-looking accused was a home-wrecker, and had got off with the police chief's wife! Hence the frightful crimes attributed to him. Somehow we sorted out the different cases and did our best to dispense justice.

Chapter 6

MY INITIATION AS A 'BUSHWHACKER'

Ultimately came the long-awaited day of embarkation for home and the long sea journey away from the sun and sand of Egypt back to our green New Zealand. We arrived at the port of Wellington, the capital city of New Zealand, on a bright sunny morning, welcomed by a band on the wharf playing appropriately 'When Johnny comes marching home again'.

After my period of welcome had expired, so to speak, I set about the business of trying to get another start in farming. This involved travelling all over the country looking at various farming propositions. In all the most desirable parts well served with roads and other amenities, prices asked for pastoral land were so high that I could see little possibility of making a financial success of the farms I was pressed to buy. Agents were making a good thing out of the demand for land occasioned by the return of young men like myself after years in the army. In their eagerness to get started again on land of their own, many paid prices that proved to be far too high, and it was later a sad business to see so many of them come to grief financially.

I concluded that the only real prospect of success lay in taking on 'unimproved' land in the less favoured and least exploited parts, and 'breaking it in' from the rough myself. Such land was relatively much cheaper. The area north of Auckland, a long peninsula stretching over 200 miles northwards into the ocean, much of it still in standing forest, seemed to have prospects. It was subtropical, with an assured rainfall and comparatively high mean temperatures, the ideal conditions for good grass growth.

After months of frustration my attention settled on a block of 1,900 acres of hill bush land, in North Auckland, most of it still in standing rain forest. This piece of land drew me in some mysterious way, and in my imagination I saw what could be made of it. I decided to buy it if I could finance myself into it on my limited capital. This meant borrowing heavily, pledging everything I owned and going in 'boots and all' as the saying is. The owner was an elderly Nova Scotian, tall, lean and somewhat dour, who had spent most of his life as a bushman, getting out the big timber trees from the forest for cutting into house timber.

He, like many others, had invested his savings in bush land he had been able to buy cheaply, and was now seeking to realise on his investment. He had come by his possessions the hard way, and when I approached him proved to be a tough negotiator. He was anxious to quit a property that was becoming a burden to him. I was keen to take it over, but neither of us dared show our true feelings in the hour or two that we bargained. It was ultimately agreed that I was to take the property over on terms that would allow me to use my capital to develop it, while most of the purchase money remained on mortgage.

So it was that soon after, I arrived in Whangarei, 110 miles north of Auckland, at that time a sleepy hollow of a small town, by Coastal boat in the middle of April 1920. I had brought my two sheepdogs with me from the South, my saddle and bridle and, in two cases, all my worldly goods. I bought a horse at a sale on the day of my arrival, and next day set out on the twenty-five-mile ride to my new home, my pace regulated to what the two dogs could manage.

It was a tiring five-hour ride, and as I reached the end of the clay road and forded an unbridged river to start on the last lap of a three mile bridle track through brambles and bush, I had time to reflect on what I had taken on. I was assailed by quite a few doubts. All my farming experience had been in the southern part of New Zealand, on the plains, and then in the high mountain tussock lands which somewhat resemble the mountain range country of the U.S.A. I knew my stuff in the management of the fine-woolled high country sheep. I had little experience of cattle. I knew nothing of the Maoris.

Yet here I was, heavily committed financially, in taking over bush land in an environment quite strange to me, where Maoris predominated and on whose labour I would have to rely to get my work done. Although slightly daunted, I was possessed by an emotional driving force sparked by the challenge that all this represented. This helped to keep me going in the ensuing years even when things looked pretty grim. In this spirit I arrived at my future dwelling, a corrugated iron hut with a bush chimney, which was no ornament to the clearing in which it stood.

That was the opening of the play. The succeeding phases of the story are told in the rest of this book. In it I hope the reader will share with me the trials, tribulations, triumphs and disasters of making a farm out of a forest. To put things in perspective I should explain the usual procedure in developing, or if you like 'exploiting' the native rain forest of New Zealand. The forests as found in the North Island of New Zealand are very mixed. There are many species of trees, but only a limited number have a value as timber. The first incursions by man therefore were in quest of these trees having a marketable value. These were often taken out years before anything was done about opening up the land for farming. Therefore I now propose to describe this work. It was of a colourful pioneering nature, and nearly always preceded the actual clearing of the land for conversion into pasture, which was of course to be my main concern.

Chapter 7

THE BUSH—GETTING OUT THE MILLABLE TIMBER

The typical rain forest of northern New Zealand clothes the hills and valleys in a dense mantle of trees of wide variety with an underscrub of smaller growth that is often nearly impenetrable. It is not unlike the jungle of Malaya. Such was the covering of most of my newly acquired 'farm'. Scattered through the forest, sometimes thinly, and sometimes thickly, according to the soil and locality, are the big timber trees of commercial value.

Pre-eminent among them is the giant Kauri of the northern forests. When sawn, this tree produces a straw-coloured fine grained durable timber that is much sought after. Perhaps next in value is the Totara, which yields salmon-pink timber of great durability. This has special value as fencing timber on farms, since it splits readily and lasts many years in the ground as posts. Then there is the Rimu, with its beautiful grain, for many years the most widely used timber in building. The Kahikatea, a large tree growing in swampy ground, produces a creamy white timber once used for butter-boxes, but now used in housebuilding since it lends itself to pressure preservation treatments on account of its softness, which allows ready penetration of the chemicals used. The Matai provides top grade flooring, the Puriri provides posts, and other trees play their part for special uses.

The first stage in bringing bush country into use was usually the extraction or 'working' of the millable timber. This was not always done. In many remote areas, without roads or other facilities, and where the owner wanted to chop the bush and get the land into income-earning grass, much good timber was burnt. How glad we would be to have it today! In buying my bush property, a condition of the deal was that I took over an agreement for the extraction of millable timber by a bush contractor then working on the property. Under this arrangement the owner of the timber, which I was to become, received a small royalty from 1s. 6d. to 2s. 6d.

per 100 superficial feet measurement on all the timber taken, this being deducted from the amount paid by the timber miller to the contractor for his work. This was the usual kind of arrangement, and as the working out of the millable timber trees was done before the remaining bush was felled for burning off, this seems to be the place to give a general account of it.

The methods of working and handling these big trees varied according to the locality, but the general proposition was always basically the same. Having felled and crosscut the trees into lengths that could be handled, the logs had to be dragged by brute strength, which was invariably supplied by teams of bullocks, to a point where they could either be loaded on to wheeled transport or floated out in a stream or river. Another method, where timber was dense enough, was to hitch the logs to a lengthy wire rope which was wound in on the large drum of a steam hauler or winch. Sometimes it was milled in a bush mill at that point. Ofter it was 'fleeted up' on the skids ready for loading on to a bush tramway, the motive power for which was supplied either by draught animals or small steam locomotives. These latter methods obviously could only be used where there was enough timber in a limited area to justify the use of expensive plant.

The bushmen that did this work were a tough capable lot—a race apart—whom no difficulty could deter, and who performed amazing feats with a few quite primitive tools. These were axes, maul, and wedges, crosscut saws, mattocks for earth work, pinching bars or crowbars, and the versatile New Zealand-made timber jack. These jacks weighed about 80 pounds and by geared cogs multiplied the man power applied to the revolving handles so that the 'spear' with its two chisel points, when applied to a log, could exert a lifting pressure of 5 tons. The only source of extra power other than man-power was usually that provided by the bullock team. This could be considerable and was effectively and cunningly applied, often using wire rope and 'snatch blocks', or large pulleys, to multiply the strain exerted. Big kauri logs commonly weighed 5, 6 or 7 tons.

Another tool which should be mentioned is the pit saw, a wedge-shaped strip of steel 7 to 9 feet in length which was used vertically by two men to saw logs into planks for dams and other uses. The log was placed across two stringers well off the ground, a line marked longitudinally on it top and bottom, and the sawyers then proceeded to saw it from end to end, one on top of the log and the other underneath it in the 'pit'. Hence the term 'pit sawn'. Millions of feet of timber were cut like this, much of it being used in the early houses.

Two other tools used were the broad-axe, with one flat side and a face 10–11 inches long used for squaring timber for stringers for bridges or dams, and the 'frow' or paling knife. In kauri bush workings it was accepted practice to select a large straight grained kauri log, and from it to split palings and shingles for the bush camp. The palings were about 6 feet long and 4–6 inches wide by ½ inch thick and were used up and down on the sides of the hut, while shingles, much shorter and smaller, ⅜ inch thick were overlaid on each other to form a sound weatherproof roof. The frow was a knife-like tool a foot long and two inches wide with a bevelled edge. It had an eye at one end for the insertion of a short handle so that when the edge was driven into the end of a billet of timber pressure could be exerted to split it. An expert could steer or guide the split, and in good timber could turn out palings, shingles and rails for yards and gates equal to sawn timber. In due course I became expert at this and made many gates, some of which are still in use forty years later. A wash drawing facing page 97 by the author shows how this tool was used. The early history of New Zealand owes much to its bushmen, who did so much with so little in resources, and were not particularly well paid.

The bush contractor was the man who inspired and organised the whole enterprise. He usually contracted with a sawmill owner to fell the trees, cross cut them into logs, and deliver the logs to some agreed point. Such a contract was invariably made at an agreed amount per hundred superficial feet measurement of the logs. A superficial foot is a piece of timber measuring 12 inches by 12 inches by one inch. A medium sized kauri log for example with a girth or circumference of 12 feet and a length of 18 feet would contain 1,944 superficial feet. All measurement was done periodically at the point of delivery, and was done by running a tape round the middle circumference of the log and measuring the length. A ready reckoner gave the content in superficial feet, from which a deduction—a 'dock'—was often made for defects in the log such as shakes—longitudinal splits—hollows, knots, etc. This business sometimes produced heated arguments as to whether the measurement was being fairly done, especially when the sawmiller was inclined to make over-substantial deductions for faults or irregularly shaped logs.

The bush gang recruited by the contractor ordinarily would include first of all a good bush cook, used to cooking over an open fire in the open fireplace of a hut, and able to bake and roast in the 'camp ovens' generally used. The cook was a key man, and fortunate was the gang with a good one. Even when his abilities with the camp oven were apt to be tempered with a disposition to go on a binge occasionally, he was not readily sacked. He was usually a master of the camp oven, in which he produced beautiful bread and scones. There would be needed one or more pairs of crosscutters. These were the men, who working as partners, did the actual felling of the trees and cutting them into lengths. They were paid on contract, so much a hundred feet on the logs cut. Other men varying in number according to the size of the job, would be needed to cut tracks for 'breaking-out' of the logs, and for forming, by hand labour, the 'skidded roads' that were so often used in getting the timber out. These were roads, usually engineered by eye, 10–12 feet in width and of easy grade, on which skids 8–9 inches in diameter and 10 feet long, were transversely spaced about five feet apart. These were carefully levelled. To transport the timber on these roads a 'catamaran' would be constructed. This consisted of two heavy hard-wearing pieces of timber, rata for preference, with up-turned ends and 20–30 feet in length. They formed runners, spaced apart by two heavy crosspieces bolted to them. On these the logs lay. This 'cat', loaded with one or more logs, with grease sloshed on the skids, enabled a team of bullocks to move surprisingly heavy loads with the minimum of effort, up to 5,000–6,000 feet at a time—say 10 tons.

The bullock team was the final and essential component of the bush job, with the 'bullocky' and his 'offsider'. The team would normally consist of up to twenty-four animals, working in pairs two to a yoke on the heavy chain that ran back to the log. An intelligent and well trained pair of 'leaders' was the key to the effectiveness of the team, supplemented by the command of language possessed by the bullocky, which could reach surprising flights. All bushmen would be capable axemen, and competent with pick, shovel, crowbar and timber jack.

The bush camp which was their temporary home was built at some central point near water in the area to be worked. It was usually of split timber palings with a roof of either shingles or plaited nikau palm fronds. A good grindstone was an essential item of equipment, axes being kept razor-sharp. The system followed in getting out millable timber varied according to its density in a given area, the availability of roads or waterways suitable for floating logs out, and many other factors.

The working of the timber on my own property was typical of much in the northern part of New Zealand. The area was remote from roads, but there were good streams which, with some artificial help, could be used for 'driving' the timber out to a larger river where the logs could be floated to mills situated on the lower reaches. In this case it was the Northern Wairoa River, a great waterway down which had floated thousands of large kauri logs to mills in the locality of Dargaville. The timber itself was scattered, involving a good deal of track-making.

The artificial help referred to consisted of the building of a sawn timber dam at the head of the main creek, which impounded a fair body of water. This was 30–35 feet in height, built of pit-sawn 9×3 inches heavy kauri planks at a narrowish part of the stream. In the lower centre section was a huge door of vertical planks perhaps 14 feet high and 20 feet wide which was water-tight when the dam was filled, but could be released outwards with a trip device, releasing the water and the logs floating in it. The technique was to augment a normal fresh in the stream by a temporarily large volume of additional water. This was expected to float and drive down the water-course the logs that had already been jacked into the stream awaiting this operation. Sometimes it did, sometimes it didn't—it could be a chancy business, often leaving logs stranded or jammed together in tricky spots. The operation of jacking the logs stranded on the banks back into the stream-bed was known as 'trimming'.

The tripping of a large dam, with many logs floating in the impounded water, could be a spectacular affair, which often drew an audience. The trip device on the downstream side of the dam had a wire attached to it from well up on the bank. At the appropriate time the man in charge pulled hard on this, and, with a roar of foaming water, the stout planks of the door were floated out horizontally on their wire loop hinges as a million or more gallons of water surged through the opening.

With it came the logs, many of them 5 or 6 tons in weight, borne irresistibly forward like corks on the commencement of their long journey to the mill. Some would grind and crash together while others tumbled end for end in this mad maelstrom that had been so carefully planned. As the dam emptied, the surge of water carried all before it as it added its impetus to a stream already running at flood level. If the 'drive' was a good one it would clean the creek out for miles, and carry the logs out into one of the main waterways. This would eventually deliver them to a point where they could be picked up for impounding within one of the floating booms adjacent to a mill.

Let's look at the overall job. The first step was to fell the trees and crosscut the millable timber into logs of a convenient size to handle. The individual logs were first 'barked' to provide a smooth surface to the log, then 'broken out' or dragged through a roughly cleared bush track to a central point or 'landing' where the skidded road started. The front end of the log was carefully 'sniped', that is, its sharp edge was rounded to allow it to skid like a sledge-runner. It was hauled by wire rope or chain bridle. This was attached sometimes by cutting a 'D' in the top half of the nose of the log, or more often by two heavy spiked 'dogs' which were driven in each side of the log a foot or two back and half-way up. At the 'landing' on the skidded road the logs were jacked, one, two, or three at a time, on to the 'cat', the bullock team was yoked up, and on the crude-looking but smoothly running sledge they were hauled to a point from which they could be conveniently jacked or dragged to the driving creek, there to await a suitable fresh. Very heavy loads could be hauled out without undue strain on the bullocks, the 'cats' gliding on the well-greased skids of a skilfully laid out road.

It was the duty of the most junior member of the outfit to grease the skids with a mop dipped in melted tallow. The 'cat' with its heavy load was braked on the down grades by means of a tail rope of flexible wire rope which was given two or three turns round a smooth stump, and paid out so that the load didn't over-run the bullocks. Occasionally there was a slip and a pile-up of logs, 'cat' and bullocks had to be sorted out.

On my property a bullocky was bringing out a heavy kauri log on a bullock waggon along a narrow side-cutting half-way up a hill face that had been cleared. The off front wheel slipped over the edge, and by the time the team was halted the top-heavy load was dangerously canted. The waggon obviously had to be pulled backwards on the road to get it on an even keel. To do this the bullocky quickly unhitched all the team except the two polers, one each side of the waggon-tongue. The team was hitched to the rear of the waggon, and the Maori 'offsider', a somewhat irresponsible youth, was instructed to guide the two polers so that the front wheels came back on to the road.

As the waggon started to move backwards the two bullocks swung the wrong way, the Maori lost his nerve, and away went the whole caboodle over the side, log, waggon, and bullocks, to crash down the hillside into the gully. On the way the fore-carriage, to the pole of which the two beasts were still yoked, broke free from the waggon. It and the bullocks landed right way up in the gully, and the terrified animals, no more than bruised, took off at a gallop, with the heavy two wheeled carriage hitting the high spots as it bounded along over a considerable cleared area. This appealed to the Maori, who, completely devoid of contrition for a disaster largely his fault, stood on the road laughing his head off as he bawled, 'By Korri—the blurry good sulky eh?!!'

Joe had the timber cutting contract on the property I had taken over, and I was always interested in his work, for more than one reason. It was all new to me, and there was no better way of learning how it was done than to watch a skilled bushman at his job. Secondly I had a small financial interest, as the owner of the standing trees, in each log that reached the delivery point. With all except kauri, this was at a 'royalty' rate of 1s. 6d. per hundred feet. This meant that on a fairly large log of kahikatea—always referred to as 'kike'—measuring 16 feet in length and 9 feet 2 inches in circumference, which contained 1,000 superficial feet, I would ultimately receive 15s. if all went well. These small sums mounted up to a figure that was not to be despised by one in my financial state!

Let's have a look at Joe and his team at work. Joe is tall and gaunt, with a slight stoop and greying hair. He has been in the bush most of his life, and knows all the tricks of the trade. He has the bushman's knack of being able to swing an 80-pound timber jack on to his shoulder in an almost effortless movement. When he drops it you can be sure it will be in the right position to either lift or turn the log that is being handled. The men are working in a small gully of damp rich soil where the moisture-loving 'kikes' thrive. They are working on two logs crosscut from one felled tree. Joe is 'sniping' the butt of the first log. This involves turning the log slowly as he chops the sharp edge of the log into a nice rounded shape on which it will skid freely over the ground. 'Hold it Tom,' says Joe. Tom rests on the handles of his jack while Bill rams the spiked heel of his jack into a fresh position. He twirls the handle until the chisel points of the spear engage firmly in the log ready to give it a further turn in alternation with Tom.

The log is nearly ready to start on its journey—this is the first phase—'breaking out'. Here comes Harry with his bullocks along a roughly cut track that leads to the 'landing' where the logs will be loaded on the 'cat'. 'Whoa—Baldy, Brindle!' and the two leaders come to a stop. Joe now proceeds to attach the heavy chain bridle to the log. To this the hauling chain will be secured. To a stout ring in a swivel, (to allow for rolling) are joined two short lengths of chain with two heavy angled spiked 'dogs' at their outer ends. Joe chops two nicks out of the sniped end of the log opposite each other half-way down, at its widest part. The two dogs are driven into the log with a maul a little behind the two nicks, in which the chain fits snugly. The ring is a foot or two ahead of the centre of the log, ready for the attachment of the hauling chain.

All this takes place in quite dense bush which surrounds the felled tree, where there is little room for movement, and now begins one of the cleverest acts of the bullock team, led by their intelligent leaders. To an accompaniment of words of command, some profanity, and a few well-placed flicks of the whip, the leaders, with the team behind them, move up to the log. Directed by the driver, they then double back and move forward into the cleared track by which they had approached. As the final pair of the sixteen bullocks reaches the log, the team is halted and the heavy hauling chain to which they are all yoked is hooked into the ring of the bridle attached to the log. A crack of the whip as Harry shouts 'Baldy! Brindle! Red! Roanie!' and the team leans into their yokes and the log starts to move. At it slides on its way, Tom and Bill sling their jacks to their shoulders and follow it, ready to jack the log clear of any stumps which may obstruct its passage.

Soon the 'landing' is reached, a set of skids at the terminus of the skidded road. Here the dogs are levered out with a pinch bar, and the team is driven back to repeat the process with the second log. Later these logs and others will be loaded on to the crude sledge or 'catamaran' and hauled over well-greased skids up a gentle incline to the top of a ridge. From there it is all downhill haulage to the 'driving creek' some distance away, and the logs, off-loaded from the 'cat', will be dragged along a well worn track to the edge of the creek. There they will lie until a heavy fall of rain seems imminent, when they will be jacked into the creek awaiting the expected rise in the water, which, augmented by the tripping of an upstream dam, is expected to start them on their journey to the mill.

There are quite a lot of chances in this operation. If the fresh in the creek is a good one, sustained by continuing heavy rain, most of the logs may be carried out and into the main river, the Northern Wairoa. When they get that far they are picked up by launches and towed inside the floating booms of the sawmills, there to await their turn to be cut into timber. Every log is readily identifiable as to ownership and measurement by brands hammered into its ends, and a number which is cut into the ends with a special type of scriber. This operation is done when measuring the logs before they leave on their long journey.

Sometimes things go wrong—the fresh in the creek may be only a short one, subsiding quickly and leaving logs stranded on the river bank for miles. Again, freak conditions in a flood, and perhaps one or two sodden heavy logs, may cause a log jam at some narrow point in the creek. I remember that for some months there was a jam like this of some seventy logs piled up in an inextricable tangle 40 feet high between two rocky outcrops that narrowed the creek. Explosives could be used at the expense of some damage to the timber in such a case, but a real 'old man' flood in the autumn ultimately swept the tangle away and cleared the bed of the stream.

The old Kauri bushmen used every artifice available to them in shifting heavy logs, including the force of gravity. Sometimes they would be rolled down a slope, sometimes shot down endways in a shute. One bushman told me of what seemed to me a terrifying procedure in which he had taken part. This was an occasion when big logs that were being rolled from the top of a ridge had to have their direction changed soon after they started to roll. 'There was I,' said Danny, 'not far from the top, with the spear of my jack partly extended, as these big things started to roll towards me I had to catch them on one end with the jack as they rolled, to start them on another slope at a different angle. As you can imagine, I had one or two narrow squeaks, and I was glad to see the end of that lot!'

Chapter 8

CHOPPING THE BUSH

Having done what was possible about milling the timber of value in his bush, the settler's first concern then was to get on with the job of chopping the bush in the first stage of converting his property into grassland. All this too often involved a visit to his local banker to arrange an overdraft. This could be a painful procedure as I found from experience. Such people seemed to me, as a young man keen to get on with the job, to be somewhat lacking in imagination, but no doubt I was prejudiced. Once he was assured of the necessary finance, he then had to find a bush contractor capable both of doing the job to his satisfaction and organising and controlling his men.

Such were not always easy to find. There was scope for a lot of roguery in these contract bush jobs, and the tricks of the trade were just as well known to the Maoris as the Pakeha.* The employer was often called on to 'stake' the contractor, that is, set him up, with a fairly stubstantial order at the local store for food, tools, and so on. You had to watch your step here—if you were too liberal, and your man was a rascal, he was quite capable of taking off with his order from the store, in which case you wouldn't see him again, and the chances of recovering anything were remote. Fortunately these types were the exception, though there were many tricks for 'getting away with it' in the actual bushfelling.

This work was done in the winter months from say May to October. The area in a contract largely depended on the finances of the settler, and very often on the success or otherwise of his interview with the bank manager. He had to think not only of the cost of felling the bush, but of subsequent outlay for grass-seed, for fencing to enclose the area, and the stock to control it. Fencing had to be thought of in delineating the area, it being desirable to have this on well defined ridges wherever possible, where it would not be damaged by slips or slumping of the soil.

* Maori name for the European.

An estimate was made of the block to be chopped, and a price per acre for the work was agreed upon. It was usual for the employer to pay the insurance premium for insuring the men against accidents. A common contract price was 35s. to 40s. per acre. The employer paid the insurance premium, which was heavy, at £6 per cent on all wages paid. The exact determination of the area was done by a surveyor on completion of the job, payment for this being shared between employer and contractor. One contractor employed by me, a Yugoslav, was concerned to find that the survey acreage was on the flat or plane basis. Since the ground was hilly he thought he had chopped a greater area than this and suspected that the surveyor and myself were conspiring to defraud him. We compromised by surveying his area in two separate pieces, which he was convinced would give him a larger return.

One's chief concern in this work was getting the job thoroughly done so that the bush felled would dry well and burn clean. If the chopping was 'slummed' enough foliage would remain green to spoil the fire, or even, in a wet season, to stop it burning at all. I knew one area not far away where the chopping was done on a 'quick and dirty' basis and several hundred acres refused to burn. It was an eyesore for many years afterwards. To try and ensure a proper job it was customary to draw up a contract specifying how the work was to be done. This provided, for example, that all the underscrub was to be cut with slashers before the main felling of the trees commenced. It set out the diameters of the different varieties of trees that could be left standing, such as 24 inches for hardwoods like Puriri to 30 inches for softwoods such as Kahikatea or white pine. To fell all these trees would have been too costly. Another item provided for would be that all trees had to be felled irrespective of size for two chains on each side of intended fence lines.

I remember going over one fence line with a real 'tough guy' Maori bush contractor by name Tungaroa. He had been in the bush all his life, was a first class axeman and would tackle anything. We came to a big kahikatea tree not less than four feet in diameter and I recall his query was 'knock him down boss?' I replied as gravely as I could 'yes knock him down Tungaroa'. He was duly 'knocked down' later entirely by axe work.

The bush contractors knew all that was involved in a contract very well, but to get them to carry it out thoroughly was another matter. I found that many of the Maoris in particular had no scruples about putting it across a Pakeha. Thus one would find all sorts of attempts to slum the most unpopular part of the job—the underscrubbing. Some of this was undoubtedly difficult, especially in dirty gullies which often contained heavy growths of gigi and supplejacks. If the employer was not 'on the job' a favourite trick was to fell the trees into these gullies so that it was impossible to inspect what was under the tangled mass. Similarly, when the chopping of the trees was begun, many trees would be chopped only to the point where they would fall but remain attached to the stump. Such trees would often remain green, refuse to dry out, and thus spoil the subsequent burn.

I now became immediately involved in this since I had to get more land cleared so that I could put on more stock to increase the financial earnings of the place. At this time (1920) bush land was being 'brought in' all over the north, and good contractors were hard to find. I was at a disadvantage in not knowing all the tricks of the trade, but my only immediate neighbours, four brothers, three of whom were returned soldiers, and all experienced in bush work, were very helpful in 'putting me wise' on many points.

The only man I could recruit for my first bush-chopping contract was a middle-aged Maori of somewhat doubtful reputation. He was of fierce aspect, and the veneer of Pakeha civilisation sat lightly upon him. He agreed to take on a contract to fell about 230 acres, and set about establishing his camp. He was a pillar of the local Maori church, a preacher, and his emblem of office was a billicock hat. This seemed to me the acme of incongruity in the environment of his bush camp, when he returned there late on a Sunday, riding a weedy pony barely able to carry his weight.

The method of felling the bush followed a generally accepted pattern. The underscrubbing was done first, perhaps thirty to forty acres at a time. A lot of this growth was tangled and springy and called for the skilful use of a slasher. If the work was well done, the cut material dried out under the larger felled trees to form a good 'carrier' for the fire when the time came to burn the chopping.

After the scrubbing, the real bush chopping commenced. By no means all the trees were chopped through . . . the skilled bushman made use of the topography and the nature of the bush he was working in. The most commonly used labour-saving practice was that of the 'drive', which was likely to be most successful on a fairly steeply sloping face. A big tree with a heavy spreading top was selected well up the hill. The idea was to start a kind of chain reaction to knock over the trees covering quite a width below this

A bush camp in the shadow of a magnificent Kauri tree, note figure at base.

Heavy labour—bullocks log-hauling.

Making ready to haul away the logs cut from a forest giant. Assuming these to average twenty-four feet in girth and ten feet in length, content of each would exceed 4,000 superficial feet.

tree, by felling it on top of them. The trees below were prepared by 'scarfing' or chopping them part way through only. While some trees, like tawa, would break and splinter readily, others, such as the koe-koe, tended to hang to the stump and remain green. A knowledge of all this was needed in preparing the drive.

All preparations having been made, and everyone cleared out of the area below, the big tree was felled, with a bellow of 'under below' as he started to tip over. There would be a huge crash followed by a succession of smaller crashes and splintering noises as the trees below broke off under the weight of the big fellow. They in turn threw their weight on the trees beneath them as the whole of the prepared area collapsed like a house of cards. The Maoris loved this sort of thing, and an accompaniment to the crashing trees would be furnished by an assortment of howls and yells from the gang.

If the preparation had been scamped the drive might be spoilt by a 'hang-up', due to some of the bigger trees not having been properly scarfed. This would then call for someone to undertake one of the most dangerous jobs in the bush, that of trying to chop through the one or two trees causing the trouble. I remember watching from a hillside the final preparations for a drive. I could hear but not see the men chopping away at the big tree that was selected to flatten the trees below. There was evidently some miscalculation, either in cutting the tree, or in establishing where the weight in the crown lay, or there may have been an adverse gust of wind. Without any warning the tree tipped and crashed *up* hill amongst the gang. There was a few seconds dead silence, then when it was realised no one was hurt, there was an excited jabber beside which the commotion of an over-turned bee-hive would have seemed mild. I have seen this happen twice; in the second case it was clearly due to a misjudgement as to where the weight in the crown of the tree was.

Maoris were seen at their best in this sort of work. They like working in a group and they are naturally good axemen. Their usual garb was a sleevless knitted bushman's singlet, and denim or palmer-nap pants. Many of them had superb physiques, which had full play here—with their muscles rippling under their brown skins they were worth seeing. A disadvantage of our push-button machine age is that today so many sons of these men, so splendidly endowed physically, are to be seen developing paunches, along with their pakeha colleagues, in sedentary jobs driving buses, bulldozers, and what have you.

Millions of acres of bush were felled, as described, over a few decades in New Zealand, tapering off practically to an end in the early nineteen thirties. This work usually came to an end in October or November, when the area of a contract would be surveyed and the men paid off. The aim was to have the chopping dried out by the hot summer sun so that it would burn readily in the early autumn.

I was soon to regret having let my bush contract to Hori. It was evident that he intended to slum the job and 'get away with it' with this new-chum pakeha—me! The first piece of underscrubbing was scamped, and I had to insist that he and his men go over it again. I was rewarded with murderous scowls, but they made some attempt to improve it. The same pattern was repeated when the chopping of the trees commenced, many being left hanging, not chopped through, on the stumps. My protests were ignored. I let them complete the first piece that had been underscrubbed, and then told the contractor that he need not start on another section until he had cleaned up the first piece, and that I would make no progress payments until that had been done. A heated altercation developed during which Hori threatened me with a slasher. He and his men tried to frighten me into compliance with their poor standard of work, but I would have none of it.

In retrospect I often thought I had taken more risks than I knew in this kind of argument. He could have felled me with his slasher in a flash, and I am sure now that the only effective restraint on his primitive reactions was the knowledge that if he 'outed' me he would not be paid. This job was a constant worry from start to finish, and gave me a very poor first impression of the Maoris. I was heartily glad when it was finished and I had seen the last of them.

This experience taught me how important could be the character and competence of the man in charge of this sort of work. The next time I employed a Maori contractor I did a bit of research into his background. He was a man called Jack Nau, a first class axeman with a small team of good men that he bossed effectively. He did such a good job, fully up to the contract specifications, that I paid him an extra 2s. 6d. an acre over the agreed price as a bonus.

Chapter 9

BURNING OFF AND SOWING GRASS

From the end of February through the month of March the bush settler took an intense interest in the weather situation. He had a large sum of money, usually borrowed, tied up in the bush that had been felled, and it was vital to his husbandry to get this cleanly burnt and the land shortly thereafter clothed in grass. He dare not burn it too early. If he did, the weeds took charge and would smother out half his valuable grass when it was safe to sow this. He dare not leave it too late, in case the early onset of the autumn rains spoilt his chances of a good burn. He tried to time it so that the chopping was sufficiently tindery to burn readily, but so that the burnt area with its valuable ashes would not lie unsown too long. Too early sowing risked the 'take' of grass—the seed might germinate but the seedlings dry off in a succeeding dry spell.

Sowing in the last week of March was usually about the optimum time, when one could expect that the rains that germinated the seed would be followed by the normal autumn rainfall. Now let's see how it's done. The decision is taken, and our man sets out, with one or two helpers, armed with kerosene torches, to light what will be a large fire indeed, involving hundreds of tons of timber. They wait till the dew has dried out and the sun is well up, and at ten or eleven o'clock the word is given 'let 'er go!' A large bush burn well under way is awe-inspiring. As the flames leap in the air and the smoke roils and billows hundreds of feet above, the heated air as it rises creates a fierce draught. At the height of the fire to its roar is added a juddering vibration in the air. It is not without some risk to the lighters. Tongues of flame 15 or 20 feet long leap into the air and on occasion may dart back at the lighter in a freak current of air. I saw one man get his tail singed like this; he made a wild leap backwards to disappear with a large splash in a deep pool in the stream bordering the chopping. His act was highly appreciated by the rest of us.

To the man who has started it all, such a fire is sometimes alarming, especially when he wonders where it will stop. If it has been properly 'back-lit' to protect fence lines, or is burning towards the green bush, he has not much to fear. There were plenty of occasions however when fires did get out of control and damaged fencing, grass, and buildings. The main burn is over in a few hours, but as night falls the burnt area is a mass of hundreds of twinkling lights as individual logs and stumps burn themselves out, to give the effect of a phantom city on the hillside. By the following morning the success or otherwise of the burn can be judged, and the settler will be elated or depressed accordingly. For years through the decades of bush land development, the autumn months were hazy with smoke as clearings were burnt off. On one occasion a fire got away from me and into an older clearing where half-rotten logs were tinder dry. As night fell the whole main ridge appeared to be aflame. It was an alarming spectacle to a worried man with visions of heavy bills for extra grass-seed and the repair of fences, but this kind of risk you had to take.

It has been well said that 'fire is a good servant but a bad master'. Fire played an indispensable part in the development of bush and scrub land, and one had to use it. Although you tried to arrange things to burn towards the green bush, or some other barrier that would check the fire, there was always some danger that it would get out of hand.

An unexpected freak wind, sometimes generated by the ascending hot gases of the fire itself, due to the topography of the surrounding hills, could cause a terrifying inferno as it roared away beyond any control. Then one could only pray for a shower, wind from another quarter, or the hastening of nightfall. During the period of which I write such a fire got away in the centre of the North Island at Rataehi, and devastated many thousands of acres.

At this point I feel I should state the settler's case for what he is doing against those who regard him as a vandal destroying 'irreplaceable natural assets', because there is so much muddled thinking about this. It is a false assumption that if all this land had been kept in forest we would have been better off. Admittedly some steep watershed areas were felled that would have been better left in bush. But, in general, it should be recognised that our high standard of living, and the exports on which it depends, come in major proportion from the millions of sheep and cattle depasturing on these grassy hillsides once clothed in bush, now some of the finest pastures in the world.

Except for the scattered larger trees of commercial value which were taken out anyway when there were means of handling them, as described earlier, the bulk of the trees of the indigenous rain forest had little or no value as timber. From the erosion point of view, a carpet of well farmed grass, along with some judicious planting of trees in vital spots, can be made to protect the soil every bit as well as the original mantle of indigenous forest. We do need some more emphasis on the necessity for such treeplanting on farms. I am all for it.

Having brought off a more or less successful burn, the next step was to get the grass-seed sown before the mineral-rich ashes were washed off, or weeds pre-empted the area. In this period of development, dealing in grass-seed was big business for the commercial firms. They made large profits, and with the more unscrupulous firms there was a good deal of dishonesty at the expense of the farmer. The current practice of having the seed mixed and bagged in the firm's store made it easy to adulterate a mixture with poor old seasons' seed which was difficult to detect. All this skulduggery by dishonest merchants had one good result. Farmers became tired of being fleeced, and pressed the Government to inaugurate a system of seed testing and certification. This was initiated and developed with great success by dedicated departmental officers, who could fairly claim that in this project New Zealand led the world. Until this came into operation the only safe thing to do was to deal with a reputable firm when buying seed.

The bush settler had no facilities for mixing the seed himself, and usually had to take delivery of the bagged seed from a horse-drawn waggon on a muddy clay road. From that point, or a shed of his own, if he had one, he had to pack the bags out to his bush burn clearing on pack-horses. This could be a lot of fun or heart-break. Road access often ended where the bags were dumped.

For my first three years I had to pack all my requirements three miles from the end of a clay road to my homestead site, crossing an unbridged river, which flooded badly, on the way. The grass-seed of course had to go farther than this, another mile or two to the area to be sown. I found I could manage up to four pack-horses, each carrying 200 pounds, single handed, but with more another man or boy was needed. At one stage two of us with ten horses were packing in a ton at a time, first with seed and then fencing wire, each horse carrying 2 hundredweights.

This is where we struck trouble—half the horses were 'brumbies' not used to a pack-saddle, and the pack-saddles, mostly borrowed, were not of a class to stand up to a buckjumper determined to dump his load. Add to this that each bag cost £7 and that the seed flowed like water each time a bag was jagged on obstructions on a narrow tortuous track, and the reader can imagine that we had a busy time. I carried a bag needle and thread in the crown of my hat, and was in the world record class for jumping off my horse, staunching and sewing up a torn bag in a matter of seconds. Arrived at the burn, the pack-loads were spread round at convenient spacing for the sowers.

In this mechanical age many people will be surprised to be told that the millions of acres of grassed hillsides they see on any trip through the North Island of New Zealand were all sown by hand. The bush burn, especially if the burn had been poor, was the next thing to a barbed wire entanglement. There were tangled heaps of timber, and logs of all sizes to clamber over or under with snags and jags to catch at the intruder.

The sowers worked in a gang, from four to eight or more according to the area. Empty seed bags were cut and sewn to form sowing bags, the style favoured most resembled a large waistcoat with two large pockets at the sides. In these each man could carry 30–40 pounds of seed, which had to be applied at the rate of about 24 pounds to the acre. Using both hands, throwing his arms wide alternately, a skilled sower covered 10–15 feet in a strip. The job was regulated by the boss, who 'made the line'. The man next to him followed slightly behind and saw that his spread of seed connected up, and so on down the line.

The boss's job was a tricky one, as not all the men would be skilled, and some were not above lightening their load by heavy sowing from a newly filled bag. He had to see that there were no misses, that all the area was covered, and that the precious seed he was paying for was made to cover the clearing at the rate decided on. Everything of course was a mass of charcoal and a more disreputable looking lot of blackfaced chimney sweeps than a gang of sowers coming in from work would have been hard to find. Some of the Maoris were very expert at this job, a good man could sow up to five acres in a day. The tangle of felled trees was such that one often had to climb through or over the unburnt heads of trees to see the ground below got its ration of seed. I employed one small elderly Maori who was a very good and conscientious sower. He travelled in a straight line, through or over whatever was in his path, and was often to be seen eight or ten feet up in the air going over rather than round a tangle of trees.

The seed as it lay on the black surface of the burnt-over area provided free meals for the birds and crickets, so that one hoped for the early onset of the autumn rains. When these came the seed germinated quickly, and a thrill of satisfactory accomplishment rewarded one as a deepening tinge of green spread gradually over what had been an unsightly charred hillside. What you were witnessing was the birth of a permanent pasture—a new addition to the world's food producing land.

The first piece of cleared bush that I burnt and sowed was a large one, about 360 acres. Although I had spread the bags of seed round carefully when packing it out on pack horses, the seed ran unaccountably short towards the end. I had to buy an extra two bags at £7 a bag when on a very tight financial budget. I had a suspicion that some seed had been stolen, but it was not until eighteen months later that I was able to prove it. It was common practice in sowing bush burns to add to the grass-seed mixture a small quantity, say one pound to the acre, of turnip or rape seed. This often germinated and grew well in ash-fertilised ground and provided first-rate stock feed. In that year I had included rape seed in my grass mixture, and I happened to know that no one else in the district had done so.

Accordingly, when I found a sample of machine dressed grass-seed containing rape seed in a shanty owned by one of the sowing gang, I was fairly certain of its origin. Under cross-examination by the police the individual concerned first came up with a cock-and-bull story about having harvested the seed himself, but ultimately confessed to the theft. He got off lightly with a small fine and the refunding of the cost of the seed. This 'long arm of the law' touch, eighteen months after the actual offence, had a very good effect on some of our local light-fingered gentry, and gave me a certain standing as an amateur detective!

There was a good deal of luck in securing good pasture establishment on a bush burn. If too long a time elapsed between burning and sowing, and there happened to be a heavy fall of rain on the ashes, there was a tremendous germination of weed seeds. Foremost amongst these were scotch thistles and inkweed, which had a smothering effect on the growth of the grass when it was sown, and could make the handling of stock on the new area quite a problem.

On one clearing millions of thistles came up in a dense mass, and we had to carry slashers when mustering the stock. The thistles also punctured the delicate membranes of the noses and mouths of lambs, inoculating them with the bug that causes 'scabby mouth'. Another weed that sometimes grew profusely was cape gooseberry. This had some advantages, for its yellow berry in green capsules made a most sought after jam. One year parties of cape gooseberry pickers on one of my new clearings supplied the whole district with this fruit for jam.

Chapter 10

THE GRAVEYARD OF THE KAURI COMES TO LIFE AGAIN

My original block of land was all hill country, without a level acre. Contigious to my front boundary was an area of 'gumland', poverty stricken stuff covered in fern, scrub, and blackberry, but of easy contour of gently rolling slopes and flats. This gumland, so-called because in the hard days of the 90's it was all dug over for kauri gum, was originally, hundreds of years ago, kauri forest. Some agency, it may have been human, or it may have been due to lightning or some other natural cause, destroyed these forests by fire. Denuded of its forest cover, the land lay desolate under an annual rainfall of about 60 inches, or 6,000 tons of water to the acre.

The various minerals providing plant food were steadily leached away by water, leaving a top soil that was nearly sterile, but able to support a scraggly growth of fern and manuka. Every square yard of it was pitted and left with a rough surface where sods had been dug out in the search for kauri gum which originated in the kauri forest once covering this land. The lumps of gum are completely inert and are unaffected by their long burial. The diggers probed for them with steel rapier-like blades 3–4 feet long equipped with spade handles. Most of the gum was close to the surface, but sometimes was buried several feet down, and every here and there we would come across big holes. So rough was some of this ground that on one part I had to get a bulldozer in to do some levelling before we could get a horse team to work on it. Big lumps of gum were usually found in the crotch of the tree's crown of branches, and upon the felling of a big tree I have seen all those present doing a scramble amongst the tangle of broken branches to retrieve what gum might be there.

When I first arrived in the north I was told that this gumland was useless and nothing much could be done with it. None the less some enthusiasts in the Department of Agriculture carried on experiments in trying to establish a pasture on this forbidding soil, and ultimately proved that it could be handled provided a certain rather costly technique was followed. This interested me, and I started negotiations with the Maori owners of the land adjoining mine. It was a protracted business, but after two years the sale was completed and I became the owner of 130 acres of poor-looking stuff which I tried to envision, not very successfully, as smiling paddocks of grass. Today this type of land is handled with heavy mechanical equipment, and large areas are still being broken in by the state.

In my time the breaking in had to be done by hand labour assisted by horses. The first job was to cut the scrub with slashers. This when dry was fired, leaving a charred area of smallish sharp stumps. As horses could not be used amongst these jags, the piece of land had to be left for varying periods according to their size until they could be knocked out with an axe. This done, and the stumps burnt, the pasture-to-be could be tackled with a three horse team and a single furrow lever plough. It was at this point that the previous history of this land began to make itself disastrously felt. Little evidence of the former kauri forest this had once been remained on the surface, but once you started to plough it, you were soon pulled up all standing as the share struck an underground root or stump. Preserved from rotting by the exclusion of air by water and a tight clay soil, these remnants of the old trees, the timber dense and heavily impregnated with preservative kauri gum, were remarkably sound. The plough had to be levered or pulled free of the obstruction, which then had to be chopped or dug out. This performance was repeated hundreds of times, more time being spent in chopping and digging than in the actual ploughing. This rough usage quickly strained the beam of the plough, which had to be straightened and strengthened by welding a second beam on to it.

Behind you the long furrows lay turned up in a ribbon of dead white clay that looked incapable of growing anything. The ploughing was a slow business, but once the soil was turned up to the air it was possible to do something with it. This land after centuries of leaching was very acid, and the essence of success in handling it was to sweeten it by fallowing and a heavy application of lime. Accordingly after winter ploughing it was left to fallow, exposed to sun and air with a little light working for twelve months. It was ultimately worked down to a seed-bed with discs and harrows in the early autumn. At this point you applied 1 to 2 tons of crushed lime-stone, and 5 hundredweights of superphosphate, to the acre, with a suitable grass and clover mixture. Although wheeled horse-drawn fertiliser spreaders were used, I was too hard up to afford one, and with one or two men spread all this stuff by hand after carting it on to the cultivation in a box sledge. This was quite hard work as you traversed back and forth over, perhaps twenty acres, spreading from a bag cut to form a large pocket holding 80–100 pounds suspended from your shoulders and waist.

I always concluded the operation by mustering up about 1,200 sheep and driving them in a tight mob around the area. This covered the seed and firmed and consolidated the ground in a way that made for good germination. The old ewes used to get very indignant over this apparently purposeless exercise, and the first time my neighbours saw me doing this they obviously thought I'd gone mad. This opinion persisted for a few weeks until the first mantle of green showed clearly and evenly over what had formerly been a piece of waste land. This was the reward for a lot of work, and I used to enjoy going out just to look at it. Once the land was sown down, the major part of the task was done, but by no means finished. The land had to be liberally treated with fertiliser and lime and properly stocked, the object being to build up the organic content of the soil till after about five years you got it into good heart.

It had to be fenced, supplied with water and drained so far as that was possible. Nearly all types of gumland soil have a very impermeable subsoil through which water moves but slowly if at all. Consequently the surface soil quickly becomes waterlogged in the winter, and one has to try and get the surplus water off by a combination of open drains and underground drains. We dug in some 12,000 field tiles at an average depth of 24–30 inches in successive winters. This was a wet heavy job which we would leave smothered in white clay from head to foot and often wet through. Today as I write I can see from my office window the rich green paddocks carrying many sheep and cattle which once seemed such a flight of imagination.

Today heavy mechanical gear has taken the hard manual work out of breaking in this type of land. Now the standing scrub is crushed and battered flat by heavy bulldozers or caterpillar tractors towing heavy water-filled rollers with cutting fins on their surface. This kills enough of the scrub to facilitate burning at a later stage.

Then heavy caterpillar tractors towing giant discs, 36 inches in diameter start on the job, with the appropriate 'set' on them having the effect of multiple disc ploughs. The kauri roots are no problem, since these huge machines ride over the top of them and in fact turn some of them up. Once the initial heavy double discing has been done, the land is fallowed, and later worked down, seeded, limed, and fertilised, as I have described.

Chapter 11

THE UPS AND DOWNS OF 'BATCHING'

My bush block was twenty-five miles from the nearest town, Whangarei, from which I had to draw my supplies. The metalled road degenerated at seventeen miles into a clay one which finally petered out after another five miles. This was as far as I could get supplies on anything with wheels. Beyond that was a largish river temperamentally inclined to flash floods, and three miles of wilderness threaded by a bridle track. From the end of the road I had to pack everything on pack-horses—if it could not be packed you had to do without it. It was amazing what you could transport on two good colonial type packsaddles on a couple of sturdy horses. A mixed load called for a lot of judgement in balancing both sides of the load evenly, the final adjustment sometimes being made by the addition of the odd brick, stone, or lump of wood in a bag on the lighter side. The loads had to be kept strapped down tight to the horse's sides to keep the weight low and avoid projections which would inevitably catch on stumps or other wayside hazards on the track.

Lighter items such as bread or groceries could be carried in a split sack across a riding saddle or in a rucksack on one's back. I and my man Friday 'batched' in a hut built of corrugated iron which had been brought in on a bullock sledge from the wrecking of a small cowshed. It measured 20 × 14 feet, with a lean-to on one side for storage of tools, and a bush fireplace six feet wide and four feet deep at one end. These bush fireplaces played a big part in the early days of settlement, and were almost invariably built of corrugated iron sheets.

They were large enough to take large chunks of firewood, thus involving the minimum work in chopping, and it was common practice to keep a 'backlog' of some suitable timber such as rata smouldering at the back. You could stand up inside the fireplace, and needed to do so, to adjust pots and pans during cooking sessions.

A stout bar hung transversely across the chimney about six feet from ground level from which depended four or five lengths of chain, each with a wire hook which could be moved up or down. The equivalent of regulating cooking temperature on the modern electric stove, was to stoke up the fire and move the appropriate pot or camp oven up or down above the flames. The key utensil in all camp cooking was the cast iron camp oven. This was no more than a heavy iron vessel usually about sixteen inches in diameter, whose sides sloped slightly outward as they rose about six inches to take a close-fitting solid lid. The bottom of the oven had three short legs. The lid was slightly convex, with a small handle at the top, but shaped at its perimeter where it fitted on to the oven to provide a channel which retained ashes on the top.

In roasting or baking, one got a good fire of say puriri or rata going, and when this got to the stage of plenty of red-hot glowing coals, you shovelled some of these on to the lid and then hung the camp oven the right distance above the fire. A good camp cook could, and did, turn out beautifully light bread, scones or 'brownie' in his camp oven. They would rival any confection produced in today's elaborate electronic contrivances, with their fascinating knobs and coloured lights, which invariably have me completely bluffed.

For slow cooking, too, the camp oven was excellent. When we were to have a roast for our evening meal, the routine was to put it on at breakast. Before leaving the hut for work I would bank the fire, cover the lid of the oven with hot ashes, and suspend the oven high where there was no risk of over-cooking. On return from work one re-kindled the fire, added potatoes and onions, put a new lot of hot ashes on the lid, and within forty minutes or so you had a delicious roast cooked to a turn. The other pots provided boiling water for tea, or enabled one to cook other vegetables or a pudding.

I did not care for 'batching' much, but since it obviously had to be done I decided to do it properly. Too many of my friends had developed ulcerated stomachs from failure to cook themselves adequate meals. For our evening meal we always had meat, potatoes and one vegetable with some kind of sweet. The latter could be rice pudding or cornflour. We always kept a supply of dry chopped wood stacked round the inside of the fireplace, so that whatever the weather we could get the fire for the evening meal going quickly. I usually reckoned to have the sort of meal described above on the table within forty-five minutes of stoking the fire. All this meant that one had to add the acquisition of culinary knowledge to an already overcrowded syllabus. I collected a few well tried recipes, my speciality being date scones, cooked as a wet dough very fast over a hot fire in the camp oven.

On one never-to-be-forgotten occasion I made a 'blue' which I have not lived down yet. My stepmother and my fiancée were to stay the night at my shanty, and a roast with assorted vegetables, cooking in the camp oven, was the central item of the evening meal. My reputation as a cook was at stake, of course, and I thought it a good idea to let my fiancée see that I knew about cooking! As they arrived I unhooked the camp oven from above the fire, and put it on the floor to turn the roast, which was browning nicely. With an inadequate fork this was not easy, and I was alleged to have assisted with the toe of my boot.

One does get into somewhat casual and short-cut methods in 'batching' as the following incident will show. Rats were always a nuisance—they used to burrow into the bush fireplace from outside and pop up from their holes in the accumulated ashes to make faces at you. I kept a loaded ·22 rifle on the wall within my reach where I sat at table, with which I took the occasional snap shot at these intruders, since all other methods of dealing with them had proved futile. One day I had two surveyors in to lunch, when one of my enemies bobbed up impertinently. I grabbed the rifle, took aim and fired. The visitors nearly jumped out of their skins. That was the end of Mr. Rat, but I realised rather belatedly that my table behaviour left something to be desired!

With no road and an unbridged river to cross, which flooded badly, one was isolated. The river was a tricky proposition, for in the autumn and winter when rains were heavy it was often difficult to judge whether it was safe to cross. I had a horse who was good in water, and when the crossing was doubtful the procedure was to take off your boots and coat and tie them to the saddle ready to swim. I could often feel the horse under me teetering on an unsecure foothold near to floating. I only had to swim once, when we fell into a scoured-out hole in the middle of the river, and I had to kick myself free of the horse as we were both swept down the river. I got out safely, but my poor horse was nearly drowned, and it took me a long time to get him back on his feet and up the bank.

I had a spill of a different kind late one dark night as I cantered along at a brisk pace after an evening visiting friends. My horse struck a black cow, completely invisible, lying athwart the track. She had started to rise, and the impact landed her on her back in a blackberry bush. The horse went down on his knees, but recovered, and I managed to stay on. The cow got the worst of it!

If I had to go to town on business or to attend a sale, I had to ride twenty-five miles, which took five hours, I stabled my horse for the night and saw that he was well fed before doing the return journey the next day. Nowadays, by car, it takes but forty-five minutes on a good road—a lot less wearing!

There were several other young men, returned soldiers like myself, who had bush properties similar to mine in the district. We decided that a party line telephone would do more to link us with the outside world than anything else. The P. & T. Department would only take a line some three and a half miles out of the town, and it was made clear to us that the remaining thirty miles of line that would be necessary would have to be built by ourselves. This was fairly typical of the attitude of far too many people in those times towards those who were opening up the backblocks. It was my experience that you had to fight hard for the various amenities over the years as they slowly became available.

Nothing daunted we set to, bought wire and insulators and hauled out totara poles from our own and other properties. A lot of the hauling and construction in roadless hilly country was difficult, but we stuck to it and got it completed. Came the great day when the telephone was installed. I met a linesman at the end of the road, having led a horse out for him. We distributed his gear in saddlebags and I packed the heavy old-fashioned Stromberg wall telephone on my back. As we set off with this lot it started to pour with rain, and my companion clearly took a dismal view of the proceedings. We ultimately arrived at my shanty as the sun came out. After a good feed we got to work, and it was a wondrous moment when, all connected up, a ring produced an answering voice literally 'from over the hills and far away'.

The telephone at that time was a godsend, and saved us a lot of time and riding, besides removing the sense of isolation and being 'by the world forgot'. Nothing was easier when, getting into a tangle with my cooking operations, I could ring central and ask a somewhat flabbergasted girl at the other end what the next move should be. The girls at central must have had a lot of laughs from the enquiries from our particular party line of bachelors.

Drawing water from a hole in a nearby swamp and bathing in

A typical bullock team in the bush.

Burning a bush-chopping—after months of work. 'The smoke roiled and billowed.'

Sheep in a new clearing—note the tangle of timber which will cause 'logstain' in their wool.

Shearing in the open—times were tough.

the creek had its inconveniences, so I decided to install a rain-water tank to supply water for cooking and washing, and a bath. Getting the bath and tank in was quite a problem as they could not be packed. They were delivered by horse-waggon four miles from their destination. I laid the 400-gallon tank on its side on the sledge, with the bath standing inside it, and with a trusty old draught horse pulling the outfit we set off for home. The bridle track which dipped steeply every so often into crossings of a creek took some negotiation. I was just preparing to pat myself on the back for bringing off a difficult transport problem, when on the last crossing, within sight of my mansion, the bottom of the tank with a loud crash collided with a concealed log I had forgotten about.

Ever afterwards one leg of the bath tended to be a bit wonky while the dent in the bottom of the tank produced an incurable leak. Henceforth we were able to draw water for cooking from a tap in a pipe led into the side of the fireplace, though it was not for some time that the bathing facilities were of the best. At first they were decidedly *'al fresco'*, the bath being situated at the brow of a hill just outside the shanty door. While this provided the best possible arrangement for drainage, the privacy left something to be desired. This I discovered one day when vigorously soaping myself I looked up to see as interested spectators the wife and daughters of the bush contractor riding up the track to his camp not far away.

In my first early days while still on my own I had an unpleasant experience. I slept in one of four rough bunks against the wall. I was awakened by a violent pain in my forehead, to find that a rat had bitten me there. I spent the rest of the night on the table in the middle of the room. This was before I had any companionship, and the transition to this kind of life, with not a soul for miles, after the good fellowship of three years of army life, was sometimes depressing. I used to look up at the forbidding face of standing bush, which seemed to tower above me, and think of all the debt I seemed to be accumulating. In such a mood I used to say to myself 'well you are a damn fool to have taken on this lot'. I stayed intermittently with a farming family some four miles away, and spent most Sundays there. They were charming to me, and the two girls in particular did much to ease the loneliness of an often worried young man, who sometimes thought he had bitten off more than he could chew. The younger one ultimately became my wife.

The girls had a pet goat, which, though tractable as a kid, was becoming a bit of a nuisance, having developed a partiality for, amongst other things, eating the washing on the line. In a rash moment I offered to carry it off on my horse and release it well back on my property where there would be other goats. This I did, not without some trouble, my horse taking a dim view of the idea. After doing this and putting in a solid day's work fencing, I returned to my shanty at dusk. The door had been left open, and hearing a noise, I looked inside. There was the goat, doing a kind of war-dance on my bed! I'd better draw a veil over immediate subsequent happenings!

Before I had a permanent 'offsider' living with me I was sometimes able to get the odd Maori to assist with various undertakings. Helping me with a fencing job, a rather fine type of middle-aged Maori stayed with me for a week. He was concerned to see a young single man 'batching' as I did, and urged on me the merits of acquiring a Maori, as against a Pakeha wife.

'Why boss,' he said, 'the wahine she stay at home—she not want to rush round and spend lots of money on clothes like the pakeha girl—she better too on the farm—she grow the kumara and the potato better'n you.' He made a good case and a true one, for the Maori women are a capable lot, and do work on the land, and in their gardens to good effect, while being good helpmeets to their menfolk.

About this time I decided I must have some help, because I was overworked and needed someone to share the batching routine. Looking back on some years of this kind of living, and to give a pointer perhaps to any considering it for the first time, I would say that batching on your own, if you are working hard, is to be avoided if possible. Shared with one or two others, if they are compatible, and willing to do their whack of the work, which should be properly organised, it need involve little hardship, and is about the cheapest way of living there is.

I was too hard up to offer much in wages, so framed an advertisement designed to interest a well-educated young fellow to whom the gaining of experience under a man willing to teach him would be attractive. I took pains to specify the type of lad I wanted, since we would have to live in close association together, and made it clear that I wanted someone of good education and family. This was evidently adjudged to be snobbish by one bogus applicant, who sent me the following amusing letter—

Remuera,
Auckland.

13/12/23

Briscoe Moore,
Mangaroa,
North Auckland.

Dear Sir,

In answer to your advertisement in this morning's *Herald* I beg to apply for the situation of 'Cadet' on your 'youthful' ewe farm. I consider I come under the heading of 'Young Fellow'. I may state that my father was a French gentleman, who arrived in New Zealand some years ago from New Caledonia in an open boat, without a swag.

I am of brilliant mental calibre and my physical development is perfect. I also am an expert performer on the Ukelele and Scottish bagpipes, while my bridge capabilities are the envy and despair of our exclusive circles.

As regards my enthusiasm for work, I fairly eat it up, and if not allowed to work twenty-four hours daily, could not guarantee to preserve my bright personality, which is one of my greatest assets.

Humour and I are bedfellows, and everytime I have appeared in Court I have reduced the magistrate and court officials to a state of helpless joviality, which was greatly to my advantage, as you can imagine.

I have slight experience in farming as my maternal grandmother was a baby farmer (never caught).

The question of salary need arouse no sordid arguments, as I would consider it a mortal insult, to be offered any remuneration at all, my only stipulation being that I will not be compelled to associate with the common farm hands after working hours.

As regards education I may mention without being accused of 'swank' that I received my education at two colleges in Australia and New Zealand vis—'The two-up College', Woolomoloo, Sydney, and 'The Fan Tan Academy', Grey Street, Auckland. I also hold a diploma, from the Auckland Correspondence College. I took the course, 'How to become a successful burglar' in six lessons.

I think the above will give you a fair idea of my qualifications and capabilities, and I hope that you will give my application your sincere consideration. If accepted I will be unable to join you for at least a week as my dinner suit is at the pressers.

Sir,
I beg to remain,
Your humble servant,
Angus McSnob.

In the event I acquired a suitable young man in the person of Ralph, not long out from England, and keen to gain experience, I did try to teach him what I could, and hope this made some contribution to his years as a successful farmer later.

In the latter stages of my batching life when I had one, and sometimes two men with me, our life became quite sophisticated. We had a bathroom, with water laid on, and the sharing of duties as to cooking and so on well arranged, so that we lived comfortably. Nevertheless there were various contretemps in the cooking from time to time. One was when the scones failed to rise. This was traced to the use of ground rice instead of baking powder! The one that hit me hardest was the loss of two camp ovens full of what should have been beautiful cape gooseberry jam. I had got this lot up to the nicely simmering stage and left Ralph, my offsider, to stir and watch the brew while I completed some urgent carpentering job in the adjoining lean-to. A burnt smell was the first intimation that the jam had gone wrong and stuck to the camp oven. We had to throw the whole lot out. What a waste!

The interests of sanitation required that we have an appropriate small movable building for an earth closet, often referred to as 'The House of Parliament' or 'The Wireless Office'. This we proceeded to build, on a pair of runners, some distance away from its intended permanent position. When completed we hitched the old draught horse to it, and one of my two men, who was a wag, sat himself on the seat with the reins in his hands ready to drive off in what he called his 'one horsepower single-seater closed model'. The comedy was completed when, near its destination, something frightened the horse, which swerved violently. This was too much for the new building on its rather narrow runners. It fell on its side and the roof came off! Tableau!

Being without a road, and completely unable to use wheeled transport, was a great handicap, and one suffered by many bush settlers. As the Government would not make available the funds for the three and a half miles of road we so badly needed, I induced others to join with me in raising a loan, repayable in rates, which drew a state subsidy, the total sum being just enough to pay for the forming of an earth road. This was before the days of bulldozers and earth-movers, and an efficient contractor with horse teams, ploughs and scoops, gave us good value for the money available.

This was after over three years of having to rely on pack-horses for transport, and it seemed to bring the outside world a lot closer, even though it was not an all weather road. Well do I remember being mystified one evening by an unusual humming noise—it was the first car coming up our road. What a thrill!

Little does the town dweller, with bitumen to his door, realise what those in the backblocks, particularly the women, had to put up with when the only road available was a clay one, or there was none at all. The forming of a road, and then, the really important thing, the metalling of it, transformed life in many ways. The wise policy of recent years in trying to give everyone metalled road access has borne fruit both sociologically and in terms of production, and very few now have to live under the conditions that were common in the '20's!

Chapter 12

ALL THESE THINGS TOO

The wise farmer, then as today, put his money, which was never enough, into productive expenditure, that is, revenue earning items. This meant grass and more grass, stock and the means to control them on the grass available. Having dealt with the technique of converting bushland to a grass pasture I must deal with the importance of fencing, often not understood, but of greater moment in this present day and age than ever before.

The bushclearing when chopped, burnt, and sown, was as a rule open to the wide—in my own case to some 10,000 acres of wild hilly bushclad land behind my property. Cattle bought out of one's dwindling funds or more often on borrowed money, could, and did, stray off unfenced clearings. Sometimes you got them back, sometimes you didn't. Sheep would wander up tracks into standing bush, but would invariably come out again—there was no feed there for them, but there was plenty of young growth that cattle liked. It was an exhausting business spending a whole day scrambling through the bush trying to trace and gather up missing beasts, often with no result.

There was therefore good reason for trying to get a fence round one's new bush clearing as promptly as one could manage. Fencing could be let as a contract, at so much a chain. A typical contract would specify that all timber (that is unburnt stumps and logs) must be cleared from the line for a 12 foot width, with posts spaced not more than 16 feet apart. All posts in dips would have to be 'footed' to hold the fence down, and battens spaced so many (usually four) to a panel. The settler would have to buy wire and staples, and pack them out on to the fence line, but all the timber for strainers, posts, and battens, was bound to be available somewhere in the bush clearing. Most of what was used was totara, though kauri and other timbers were used for battens. The fencers had to split the material, haul it on to the line, erect it and strain the wires, normally seven in number.

Sometimes the settler himself with one or two men on wages, did his own fencing. I erected several miles of my own fencing like this, and took a pride and satisfaction in this constructive work as one saw each newly completed section behind one with the rays of the morning sun glistening on the new tightly strung galvanised wire. It was a very satisfying job. In splitting the larger logs for posts and battens we used blasting powder to open and quarter the logs. These charges made a good bang with a cloud of smoke. This operation had a special appeal for one of my dogs called Darkie. As I lit the fuse and ducked back to a safe spot, Darkie's ears would prick in anticipation, and when the bang came he would dash forward towards the log with a series of approving barks. On one occasion I put in two charges, one to split the log and the other to blow an unwanted knot off the side. I lit both fuses and did my usual sprint. There was a loud bang as one charge went off, and Darkie dashed forward. Then the second charge exploded and Darkie pulled up all standing as a large chunk of timber whizzed past him. I never saw a more surprised dog.

Apart from restraining stock from wandering off into the bush, fencing the bush clearings into enclosures of appropriate size was very necessary to enable adequate control of the grass by stock. The technique was, and is today, to eat the grass out in the different areas in succession by heavy concentrations of both sheep and cattle, and then close them up to allow the growth of the new short protein-rich grass. This is known as rotational grazing. When properly practised it gives the highest carrying capacity, and by forcing the grass plants to root strongly it gives a dense strong-growing sward. Today, with larger amounts than ever of phosphatic fertilisers spread on our pastoral land, mostly from the air, the importance of subdivisional fencing and control of stock on the increased grass growth is being recognised.

In the early days slips due to torrential rains gave us a lot of work. As the roots of the one-time forest trees loosened their hold as they rotted away, slips large and small appeared all over the place after a heavy rainfall. Wherever possible we sited our fence-lines on ridges, but in many places they had to run across sloping faces. These were the places where the slumping and slipping of the soil wrecked our new fences. Huge lumps of waterlogged soil, perhaps a hundred tons in weight, slid down the slope taking everything with them, and making large gaps in the fencing. The soil in these slips was always of a glutinous consistency, which nearly sucked your boots off, and the repair of this sort of damage was exhausting work. The last straw was to come home after a day of this, to learn that the road was also blocked by slips. Despite the scarred appearance of the hillsides for some years after this sort of visitation, these bare patches, with the assistance of some grass-seed, ultimately grassed over, and there is little to be seen of them today.

Mud! How I hated it!

Coming, as I had done, from a dry part of the country, I never got used to the mud. We had a clay subsoil, and in the autumn and winter months of heavy rainfall, when fencing, draining, erecting yards, or just digging, the mud was literally everywhere. Several pounds of it stuck to your boots, it stuck to your spade, axe or rammer and made the handles greasy and difficult to hold. It became ingrained in your hands, covered your clothes, and even got in your hair. You slipped and slid everywhere, and so did your horse. He occasionally fell over with you on the slippery hillsides. I became 'fed-up to the back teeth' with it every winter, and it was a relief when things started to dry up about September.

Facilities for handling stock, in the form of cattle and sheep yards were a necessity. We built these ourselves, entirely from split timber, using totara for posts and either totara or kauri for rails. To start with we had to draft our cattle in two primitive enclosures formed by long thin kauri rickers laid one on the other at the corners as in a log house. Doing this on foot one day with a mob of lively steers, I took the cleanest knock-out blow I have ever had on the point of the chin from a well-aimed kick from a steer's hind feet. My offsider dragged me from the yard, and I came to just as he was about to douse my recumbent form with a kerosene tin full of water.

One thing we had to build was a sheep dip. This was a concrete bath of the pot-dip type, with a lead out to concrete draining pens. Above the bath was a pen 6 × 4 feet built round a platform set on an off-centre axle so that the weight of the seven sheep it held was mostly on the side towards the bath. On releasing a trip, the platform tipped sideways and shot its surprised occupants into the dip. A small netted pen beyond the platform held two decoy sheep that were supposed to entice each fresh lot into the trick pen, but while the lambs ran into the pen freely, the older ewes were not to be bluffed and took a lot of forceable persuasion. Do sheep remember? Ask anyone who has had to dip several hundred old ewes in a dip they've previously been through!

I had the opportunity of experiencing the effectiveness of this device myself one day when dipping sheep with the help of a young Maori. Getting on to the tip platform to catch a sheep for a decoy, I accidentally bumped the trip lever. It worked perfectly, and I and seven sheep shot into the bath of murky and ripe-smelling dip. My Maori helper was dying to laugh, but wasn't sure it was safe to do so until the humour of the situation had struck the victim and we both let ourselves go.

There was always more than enough to do in development work, apart from the current routine work in caring for stock. There were tracks to be cut and rough bridges to be built. Farm buildings had to be thought of, especially a woolshed in which sheep could be held under cover and kept dry for shearing. In the early days I had to shear my sheep in the open, a nightmare business in a showery wet climate. I had two blade shearers and one who shore with a machine. This was operated by a small engine mounted on top of a heavy post, and was made by a well known firm. It was quite the most unreliable mechanical monstrosity I have had to operate. It would run perfectly as long as I was nearby. Then would come the time to muster and bring in more sheep. Off I would go with one ear cocked backwards listening to the engine's explosive racket. When nearly out of earshot there would be a loud bang and the thing would stop dead, whereupon I would gallop back to wrestle with it.

This leads me to the reflection that I have had many similar experiences with unreliable machinery, nearly always due to two causes. Either the fundamental design was bad, or the assembly done in New Zealand was at fault. Far too many English machines, such as mowers, showed evidence of amateur engineering and no understanding of the effects of vibration or use on rough ground. The American gear was far better, and where a castellated nut with a split pin was indicated, so that it would not rattle off, then it was provided. Many a time I wished that the so-called engineer responsible for design of some machine had been with me, so that I could have rubbed his nose in the failure of his apparatus.

Before we had a road, the only way of getting the shorn wool out was to sledge it out, three bales to a sledge. One year I walked the wool out on the dry sheep, shearing them at a shed some ten miles away. This wasn't an unqualified success, since we got caught with bad weather. A temporary solution to my shearing problem came with the building of a woolshed by my only immediate neighbour. For some years I was able to shear in his shed and pay him rent for it until I could afford to build a shed myself.

As the farm was developed and increased in carrying capacity, expenditure on improvements seemed to be never-ending. One seemed to be permanently 'broke' so far as ready cash was concerned. A shed in which to milk the house-cow, and in which we could store saddles and gear used daily, had to be built. Then as tools and implements were bought, an implement shed was needed to house them—they are far too valuable to leave in the open in our 65-inch rainfall. It is surprising the time and timber it takes to build a few weather-proof dog-kennels.

When we started to bale hay, instead of handling it loose in stacks, a hay barn was needed, and about the same time we had to build permanent quarters for shearers and men working on the place. The latter type of buildings have to comply with standards enforced by the Labour Department and are a costly item in equipping a farm. On most gumland areas good water for stock is scarce and the land has to be reticulated. This means a pumping plant, long pipelines, and cisterns piped to troughs. In the winter months there was tree-planting to be done for shelter-belts and timber production, with fruit trees for the orchard.

A wet weather job was the making of gates for new fences, and the repairing of those the bulls had smashed, this being one of their favourite pastimes. With the high rainfall of our northern region there were always drainage jobs to get rid of surplus water. This usually meant either digging open surface drains, or putting in underground drains of four inch earthenware tiles at depths of about 30 inches—a wet muddy job.

Chapter 13

BACKBLOCKS SHEPHERD

In the main, New Zealand's overseas revenue on which we depend for all our import requirements and the servicing of our overseas debt comes from the products of our livestock industries. These are wool, meat and meat products, dairy produce, hides, skins, tallow, and some oddments. The livestock farmer, which most of us are, is therefore of some importance to the country's economy. On his ability to provide good pasture and utilise the grass from it with maximum efficiency in breeding growing and fattening stock depends some 93 per cent of our export output.

When I took over my farm I had some 300-odd dry sheep. I had then to set to and buy ewes and beef type cows as a nucleus of a breeding flock of Romney sheep and a breeding herd of Hereford cattle. As a matter of interest the farm today forty-five years later now carries 3,500 sheep and 650 Hereford cattle, all bred on the property. This is not exceptional—it is matched by hundreds of other farms and is an indication of the farmers' contribution to the country's progress.

As many thousands of acres of bush land were being 'brought in', as the phrase has it, in the years immediately following World War I, the demand for stock was keen, and one had to buy what one could where one could. This was long before the days of trucking stock, and getting them home often meant two or three days droving on the road. This was very wearisome, with sheep especially, the last piece being the worst. I had to ford or swim them over a fair-sized river, and work them up a narrow bridle track for three and a half miles amongst masses of blackberry with its fiendish tentacles.

From what were often unpromising foundation stock, one had to use all one's knowledge of breeding to gradually evolve something much better. This was a long slow business with no cheap short cuts, and took many years. The essence of it all was first to have clearly in your mind what your ideal animal was, and to start in that direction by buying the best sires, in rams and bulls, that you could afford, a costly business. Thereafter on the female side one had to cull the flock or herd as ruthlessly as possible. Since new land was being broken in all the time, with increasing numbers of stock needed, this was not easy.

In my sheep flock I culled for age and defects of size, conformation, wool type and so on, but above all for constitution. What I tried to do was to build up an even flock of good quality sheep carrying a desirable fleece, and because of good constitution having the ability to 'do' well even under hard conditions.

Over the years all this had its effect on both sheep and cattle, to the point where surplus stock sold from the farm today meet a good demand. This breeding of good stock is a major factor in the value of our exports. Even racehorses figure in this, and we continue to annoy our Aussie cousins by winning their principal races. Our type of farming can never be 'collectivised' or mechanised beyond a certain point. It depends for its success very largely on the highly individual skill of the farmer, and will continue to do so.

In the fifteen years following 1920, there were violent ups and downs in the markets for our primary produce, culminating in the disastrous slump of the 30's, of which more later. My first wool clip was seven bales, which when sold by auction in Auckland brought $3\frac{3}{4}$d., one quarter of the value of the same type of wool the previous season. This was in the minor but serious enough slump of 1921. What stuck in my gullet was that the woolbroker had the nerve to say that in the circumstances he thought it a good price! Similarly with stock—I remember selling my surplus ewes one year for 25s. and being forced to take 2s. 10d. a head for them the following year as the big slump began.

I shall never forget that sale, which took place at a crisis in my financial affairs. At the poor prices offering, there was little incentive to prepare the sheep properly for sale. I accordingly offered my 250 surplus aged ewes in one lot, and they were sold in a largish yard. As the auctioneer called for bids, one wretched old ewe, quite the worst in the mob, paraded herself round the outside of it, with disastrous effect on the bidding. I learnt my lesson, and from then on always took out the worst sheep in the mob, and graded the rest according to condition and quality so that the buyer had the choice of even lines.

Getting sheep into the market involved two and a half days droving, starting in the cool of dawn but running into the hot dusty afternoon hours, which dragged interminably in dust and sweat as one coaxed tired sheep over the last few miles. On one occasion we had unusually heavy traffic through the mob of ewes we were driving, each vehicle having to be cleared through the sheep by one of us. My school-boy son counted these, and at one stage remarked 'thirty-four so far, but only two thank-you's'. One ultimately got the sheep penned up in the yards, but it was an anxious time until they were put up to auction. You were never sure whether you would get a payable price, or be forced to make a decision between taking a price that didn't cover costs, or the alternative of another two and a half days driving tired sheep and dogs home again in the hope of a better price later. In practice you were usually so tired and fed-up that you took the best offer you could get.

As more land was developed an increasing proportion of time went in looking after stock, in all the routine work. Broadly this started with the mating of the ewe flock with the rams, 'flushing' them on good feed to ensure a good conception rate, and then all the subsequent work up till the lambs were weaned nine months later. The sheep all had to be crutched in the early winter, clearing the wool from under the tail and away from the teats both to stop them getting 'daggy' and to give the lamb when it arrived good access to the milk supply.

Stock disease made itself felt, particularly with footrot in the adult sheep, and parasitic internal worms in the lambs being reared as replacements in the flock. This involved much work. External parasites were dealt with by dipping in a standard preparation in early autumn, which was fully effective, and internal parasites by dosing with various specifics. In this work every individual sheep had to be handled.

Ultimately lambing time came round and one was fully occupied in the job of sheep midwife, picking up cast ewes and helping the old girls to overcome all the abnormalities and difficulties they encountered, such as wrong presentations and over-large lambs. I never ceased to marvel how the complex process of birth of the young animal came off successfully and without assistance in the vast majority of cases. When one could save a ewe which by herself could not deliver an oversized lamb, or straightening out the tangle of limbs in the wrong presentation, one felt a real glow. It is a satisfying but exacting job, the most heart-breaking part being when a night of bitter wind and rain wipes out dozens or scores of new arrivals in a hostile world.

Come with me now on the lambing round on this dull showery

August day. We have about 1,200 ewes to get round, spread over six separate hill blocks. My horse Jumbo is a steady old character, who will stand when I drop the reins over his head. Spark the dog, my other assistant, is full of bounce but knows how to hold up a ewe so I can catch her. One saddle-bag contains my binoculars. These save a lot of riding in enabling one to spot sheep in trouble at a distance. The other saddle-bag contains a bottle of disinfectant, a couple of plastic devices for use with 'bearing trouble', a length of stout twine, and a tin containing tubes of penicillin. The last item is a sheath knife, to be used in the last resort in hopeless cases. Strapped across the pommel of my saddle is a clean chaff sack to contain any wool that has to be 'plucked' off dead sheep. This is the shepherd's basic equipment.

Off we go, up one of the bulldozer roads into the 'Dip' block. These ewes are halfway through lambing, and here are a lot of new arrivals this morning, most of them being assiduously licked and nuzzled by their proud mothers. The maternal instinct of the Romney ewe is a strong one, much more so than that of the mountain sheep of the South Island where I did my first shepherding. They are splendid mothers, and it is only with the occasional young ewe that has had a bad time that we have trouble in getting her to 'take' to her lamb.

We reach the top of a spur from which I can get a good look round with the binoculars. All seems well except for a white patch showing above a small hollow where a tree has been uprooted in a distant siding. 'Looks like a cast ewe,' I say to myself. 'We'll go and have a look.' Sure enough, when we arrive, there she is, a bulky ewe weighed down by her wet fleece, kicking vainly in an effort to rise in the small hollow in which they so often get themselves trapped. I turn her over, work and rub her legs to get the circulation back into them and then hold her up on her feet. A few staggering steps and over she goes again. We repeat the process several times until at last she can stand on her own, though still a bit wonky. She'll probably be all right now, but I mark the spot and have another look at her later.

On we go into the next block, where I quickly spot a ewe with 'bearing trouble' as we call it, or eversion of the vagina, a displacement typical of the lazy fattish ewe. This one will take some catching. Spark helps me to corner her amongst some timber in a small dip, and with a flying tackle I grab her and tip her on her side. First step is to relieve the pressure on the distended organ by manipulating it to empty the bladder of urine. This done, it goes slack like a punctured football, and holding up her hinder end I work it back into its normal position. The T-shaped plastic device is inserted to hold it in position, the ends of the top of the T being tied to the wool each side of her tail. Then I let her go and hope for the best. She will need checking daily until ready to lamb. Some of these we save and lamb successfully, but others prove to be chronic cases and ultimately have to be killed.

A bit further on and more trouble appears, in the form of a ewe running around with the swollen head of a lamb protruding from her rear end. Spark helps me to bring off another 'All Black' tackle. This is a case of mal-presentation of the lamb. The forelegs, which should come with the head in normal birth are back in an abnormal position and the ewe is unable to push the lamb out because the shoulders are jammed in the narrow passage. The remedy is to push the head and neck back and fish for the forelegs with my fingers. This I do and after a struggle succeed in getting both fore-feet out alongside the head. Then with a gentle pull and downward pressure the lamb comes away and a normal birth is safely accomplished. In spite of the swollen head the lamb is still alive and will I know soon return to normal. I prop the maternity patient up in a comfortable position and rub her lamb over her nose so that she gets the scent of it. Then I duck away from her and get the hell out of it before she becomes alarmed and clears out.

So the round goes on, with a sharp eye for anything in trouble. I find another ewe that needs assistance in delivering her lamb, another one cast flat on her back, and one tied up in a blackberry bush. Sometimes the length of twine is useful. Looped at the end it can be worked round a foot and used to pull the leg of the lamb forward to its proper position for delivery in cases of mal-presentation.

A common problem is the wet shivering abandoned lamb, the tiny waif whose feeble bleat and empty tummy arouses one's compassion as it wobbles towards you on shaky legs. It may be the first offspring of some irresponsible young ewe lacking the maternal instinct, or the weaker lamb of twins whose impatient mother won't be bothered with it. Here is one, and I remember a forlorn mother bleating over her dead lamb that I had seen earlier. Picking up the orphan, I locate the bereaved mother, and, collecting her dead lamb, work her quietly into a mothering pen in the corner of the block. There I quickly skin the dead lamb in shepherd's style, that is, cutting the skin at knees, hocks, and round the neck, and pulling it off in tubular form. After a drink the orphan is fitted with the skin of the dead lamb. I squirt some milk on to the skin and rub the newcomer over the ewe's nose. She sniffs at it with interest—'Not Tommy come back to life surely?' But that's how it is as far as she's concerned, and the chances are that when I visit the pair tomorrow to remove the skin the foster parent will have accepted her adopted lamb for good. This 'mothering-up' of odd lambs takes quite a lot of the shepherd's time. As a last resort, if no foster-mothers are available, the stray lamb often ends up as a child's pet.

The pre-lambing shearing of ewes in the colder months of the year has drawn a good deal of criticism as a practice that is thought to be unduly hard on the sheep. The practical advantages of it are not so well known. These are, firstly, that the ewes lamb with less trouble and do not get cast on their backs or sides as so many do when carrying a heavier fleece; secondly it avoids the necessity for dragging hundreds of lambs into the yards at shearing time, where they spend hours, getting hungrier and hungrier while their mothers are shorn. Thirdly, the early shorn ewes take their lambs into cover in rough weather, whereas the full-woolled ewe, comfortable herself, will stay on an exposed hillside with her new-born lamb. The individual farmer has to decide which practice fits his own circumstances best.

As a young man, I was shepherd on a large property owned by a Scottish land company with headquarters in Edinburgh. It was the custom to cable them about momentous happenings. A southerly blizzard with bitter weather resulted in heavy loss of lambs in the middle of lambing, which was duly reported. The Edinburgh manager being on holiday, his assistant stepped into the breach, and the astonished station manager received a cable which read 'suspend lambing operations till weather moderates!' Unfortunately it isn't as easy as this, and the livestock farmer has to put up with the slings and arrows of misfortune of this kind quite beyond his control.

As the lambs start to grow in the spring, comes the time for 'docking', that is, marking their ears for identification, cutting off their long tails, and castrating the male lambs. Mustering a couple of hundred ewes with their young lambs calls for a lot of patience. In the final stages, as the mob comes together approaching the yards, many of the youngsters have temporarily lost their mothers, and the air is loud with the bawling of the ewes and the bleating

The bush shanty that was the author's home for the first five years. The bullock team has just hauled out the rimu log to be cut into timber for a proper homestead.

Mustering in a large mob in the tussock high country of the South Island of New Zealand.

Handling sheep on the much lower hills of the North Island.

A Fletcher aircraft spreading its lifegiving load of phosphate on pastoral hill country.

of the lambs. It is at this point a 'break' of lambs is likely to take place, thirty or forty bolting together back towards their haunts, regardless of dogs or men and scattering all over the place. Good dogs, patience and know-how are needed to recover the truants. Once yarded, the shepherds breathe several sighs of relief, and the usual formula is to boil the billy for a mug of tea before getting on with the job. These musters for docking take place in the spring, and are often timed to coincide with the school holidays. Farm children love any work involving lambs, and they can be a considerable help.

They invariably have pet lambs and calves of their own to rear, and this kind of association and learning to care for animals is a feature of farm life much to be encouraged. When the lambs are tailed the tails are piled in two heaps according to sex. Every sheep farmer hopes for a good percentage of lambs—all his efforts of management are devoted to this end, for the larger the number of lambs the better the income. As the tails are tallied on completion, the percentage, that is the proportion of lambs to total ewes, is quickly worked out. With good shepherding and reasonable weather the percentage may be rewarding. It may easily be disappointing if, in spite of good care bad weather has decimated the newborn lambs.

A similar job has to be done with the young calves, and this was one where we had to watch our step, for an enraged Hereford cow that thinks you are trying to pinch her baby is not to be argued with, and is far more formidable than a bull. In the early days all the Herefords were horned, and we had a few exciting rodeos. On one occasion a horse was gored in the behind, and on another, one of my men had the seat of his pants neatly taken out.

The big event of the year on any sheep farm is shearing time. It is a time of long hours, heavy work, and if the weather is threatening, as it usually is, some anxiety. The different lots of ewes and lambs have to be mustered and driven in to the woolshed. There the lambs are drafted off into one yard while the mothers are shedded ready for the shearers. The object is to get them through, back to their lambs, and out to their pasture, as quickly as possible before they become too hungry. That way they 'mother up' quickly and the ewes are not so liable to be affected by weather. The main protection the shepherd can give his ewes after shearing is a full belly. It is the hungry empty sheep kept too long about the shearing shed that succumbs to a cold snap.

In a showery, fickle climate, getting the sheep with their full fleece under cover dry, is one problem, and getting them out and back to the pastures they know, under good conditions, is another. A mob of shorn ewes with their lambs delivered back to their grazing block can be a worrying sight to the unsophisticated. The ewes' main concern is to top up their stomachs, and off they go, apparently disregarding the mournful bleating of the lambs, who are thirsty. It looks a proper mess, as though most of the lambs may lose their mothers permanently, with ewes and lambs scattered singly and in groups all over the place. Not so however—the maternal instinct, one of the strongest in nature, soon asserts itself.

At first in ones and twos, and then in increasing numbers, the ewes, their immediate pangs of hunger satisfied, come back to look for their lambs. Watch this one, as, with an inquisitive bleat, she visits in turn several disconsolate lambs, literally turning up her nose as the smell of each proves to be wrong. This does not stop one bold lamb from trying to pinch a feed, for which he gets knocked endways with a good butt from the ewe. So the quest goes on until one lamb pricks up his ears at the sound of a bleat he knows, and runs to a joyful reunion with his mother. She nuzzles him fondly as he indulges in a long drink with an ecstatically wriggling tail.

There is immense satisfaction and sense of accomplishment to reward the farmer in all this, which, to any thoughtful man, offsets the often disappointing financial results of farming, especially when hit by cataclysms of nature such as floods or droughts.

The advent of radio, with weather forecasts throughout the day, is of great help to the sheepfarmer at shearing time. If he knows an anti-cyclone is moving on to the country he'll plan to push the ewes with lambs through the shed while the weather is good. Conversely if bad weather is forecast he will be likely to switch to dry sheep which in emergency can be put back in the shed after shearing if necessary. This can't be done with ewes and lambs. I composed some doggerel which envisaged the sheep-farmer huddled over his radio to get the forecast, the first verse of which went:

'With anxious mien he twirls the knob,
And thinks about this awful job
Of trying to shed the sheep while dry
In weather fit to make one cry.'

So it went on until he leaves the set with a large smile upon learning of the advent of fine weather!

For many years our shearing was done by Maori gangs of six men—three shearers and three shed hands. They would be engaged beforehand to start on a certain date. Sheep would be mustered and shedded ready for them the night before. This was always a time of acute worry—would they turn up to start in the morning? During the night we would count the arrivals by the hoofbeats of their horses, and as the bell was rung just after five o'clock for the early morning cup of tea, we would anxiously scan the team to see who was missing.

If it was one of the shearers, and we had the shed full of wet ewes, with a mob of lambs bleating their heads off standing in the yards outside, things became complicated, and someone had to dash off to the Maori settlement six miles away for a replacement. Frequently the missing man would turn up during the day, or the next night, quite unconcerned. Some of the Maoris were reliable, but many were not, and this irresponsible attitude made them difficult to employ. The worst of them came to be regarded as unemployable.

The Maori shearer seemed to have a natural aptitude for the work, helped by his often superb physique. Take a look at the gang in operation. On stand one is Bill, six feet, and splendidly proportioned, driving his hand piece unhurriedly but surely through the wool as it falls away from him in a creamy mass. He does a consistently good job. Next to him is Eru, a good shearer whose overlarge stomach tends to get in his way. His pants are precariously suspended by a twine belt round the lower part of this, and manage somehow to defy the laws of gravity. At the far end of the board is Pera, of middle height, solidly muscled, and tough as old boots. He can go on for hour after hour. He farms his own land and does so well. He is also a good axeman.

Tahi the 'fleeco', an agile eighteen, is kept on the hop sweeping the board, picking up the bellies and gathering the fleece from each shearer as the last of it is peeled from the sheep. This he throws, with some skill, on to the wool table which it covers, spread out with the skin side down. Hori works with me on the wool table, skirting the fleece and rolling it ready for classing and baling. Jack Waa, on older man, is the presser, who bales up the wool as the different bins become filled with the soft lustrous fleeces.

Things are going well, and one of the shearers starts to sing—

this is taken up by the others, and with their sense of melody it is delightful to hear. This it occurs to me, is work as it should be done.

My wife has a strenuous time feeding these men, which she does very well—three main meals and three teas, starting with one at 5.10 a.m. Often a 'hanger-in' or 'spare part' will turn up with them for a meal—we are never quite certain who he is or what he is there for, but he gets fed just the same! The men work in 'runs' none of which exceeds one and three-quarter hours—a work day totalling eight and three-quarter hours. Quite often, after a strenuous day, we will see flares down at the creek, where they are fishing for eels, just to round out the day's entertainment. They cook these in the fireplace in their quarters, leaving a decidedly fishy aroma about the place when they depart.

The next items in the sheep year are those of weaning, lamb shearing and dipping. It is at this point that the fat lambs for export are drafted off. They provide a major amount of revenue on most sheep farms. Today it is a simple enough business—the stock truck backs up to the farm loading race, the lambs are loaded on, and in a matter of an hour or two they are delivered at the works.

In the first difficult years getting a mob of 200 or 300 lambs to the works was quite an undertaking for two men. Newly weaned lambs are mad little things, difficult to handle at any time, but when they had to be driven on bridle paths through blackberry flats and then on tortuous roads with four wire barbed fences it was no fun. We had three days of this, and I had visions of half the lambs being rejected for bruises as they went backwards and forwards under the bottom barb of those wretched fences. Once we arrived at the rail station where they were to be trucked on the afternoon of the third day, where the final straw for the camel's back was awaiting us. The trucks which had been ordered were there, but locked on a siding where we could not get them to the loading yards.

This was too much for me. I improvised a crowbar with which I wrenched off the padlock holding the block over the rail. This delayed us, but we got the last truck loaded as the train arrived, and never was I more glad to see the end of a journey. This incident had a sequel soon after, when I was asked to give evidence before a railway commission. I recounted this happening, upon which one member, a stuffy departmental type, pounced on me with 'Are you aware that is an indictable offence?'

'That may be so,' I said with some heat, 'but what the hell would you have done—miss the train and space booked at the works after three days on the road?' This upset the second man on the commission—he burst into laughter and I was let off with a warning.

To get back to the business of taking the lambs from their mothers, there is a tremendous bleating and bawling as the ewes are finally weaned from their lambs. It goes on for several days, when both lots settle down to a separate existence.

Shearing the lambs is a straightforward job, and, divested of their fleeces they are an acrobatic and fast moving lot that challenge the dogs with their speed, and take some handling. Then comes the dipping, to kill any external parasites on the sheep, and by leaving a residue of toxic material in the fleece to give them some protection against reinfestation by ticks or lice. With the use of new drugs having greater ability to penetrate the fleece, the plunge type of dip in which the sheep were completely immersed has to some extent been superseded by various types of shower and spray outfits. These involve much less work, but it is doubtful whether they are as effective.

The final work of the sheep year before the whole cycle recommences is that of sorting out the surplus sheep for sale, which is done in January and February. At this point the young sheep that have been reared, mostly young ewes, take their place in the flock, while the old ewes, usually at the age of five years, are taken out and sold. These are eagerly sought by the fat lamb farmer, who renews his 'flying flock' every year or two. The old ewes are run under good conditions, and are bred to a ram suitable for fat lamb production. At five years of age they give a good percentage of lambs, and after one or two years are fattened and turned off to the freezing works. Most of them are now absorbed by the Japanese market. At the same time inferior sheep of any age are sorted out for sale, along with surplus wether lambs and small 'cull' lambs. These latter are often bought by dairy farmers, and run under good conditions which transform them into surprisingly good sheep. This then is a brief survey of the cyclical work on a sheep farm every year so far as the sheep are concerned.

The beef cattle which are run in association with sheep on a fairly large scale in the North Island of New Zealand take a good deal less work than the sheep, since they are not shorn, though a good deal could be said for the evolution of a wool-bearing cow, something on the lines of a super yak. Calves which are reared for replacements are subject to the various ills of all young stock, and have to be carefully tended, watched for signs of parasitic infestation, and given the pick of the feed. In general, the cattle are used to control the feed and to clean up rough stuff of no feed value to sheep. They are the hill country mowers, and a fine balance in judgement is called for in controlling the pasture to the right degree without damaging the cattle in the process.

Not so many years ago cattle were merely a sideline on a farm such as ours. Prices for the surplus cattle sold were low, and their chief value was as hill pasture implements that enabled their owner to keep the grass in the best order for the main revenue earners, the sheep. In recent years this has altered completely. Rising standards of living, and the development of new markets, particularly in the U.S.A., have created a brisk demand for beef products at good prices. While we still make the traditional use of cattle in pasture management, the surplus for sale now make a substantial contribution to farm revenue, and we are building up our beef herds in both numbers and quality. The town consumer in New Zealand sometimes complains about the high price of beef—he or she needs reminding that they are still getting the best and cheapest meat in the world.

The horse has been, and will long remain a faithful friend and servant of the man on hill country, and it is far too readily assumed that his day is done. We shall always need him on the hill country that is too steep for wheeled transport, which still carries the greater part of our stock of sheep and run cattle. I look back with fond memories of some old friends. There was 'Jake' my bay gelding that carried me for years. Affectionate and reliable, one could drop his reins anywhere and be sure that he would not clear out and leave you while you attended a sick sheep. Then there was old 'Toki' the first draught horse I owned. He would haul anything in reason, but, if overloaded, his ears would twitch, he would turn his head to inspect the load, and with a disapproving expression make it quite clear that it had to be lightened before he could be expected to do any more.

Then there was 'Dolly' and her son 'Baldy' bred on the place. Together they made a splendid team in the sledge or when used for hauling hundreds of tons of waste timber into heaps for burning. Baldy had one aversion—he was allergic to aeroplanes. They caused him to gallop about madly and crash through fences. If aircraft were working from our strip we had to see that he was safely parked some distance away. Before the days of tractors, horses provided the sole means of transport on the farm—as pack-horses on the steeper country, or yoked to a sledge, which carried half a ton, on the easier land.

The traditional sledge was built from two straight pieces of white tea-tree or manuka cut out by the roots so that the flange of the root made the curved upturn on the front of the runner. Three crosspieces pegged to the runners with heavy pegs, and some decking of split timber completed the outfit. It was flexible, and on striking a stump the runner would give instead of smashing something as would be likely with a rigid structure. Later we shaped solid timber runners and shod them with steel shoes. They were reversible end for end and stood a lot of wear.

I have said little about dogs, without which the sheepfarmer could not do his work. Fortunate is the man with a good team. They are difficult animals to train and work with, demanding a particular type of patience that not everyone possesses. I myself am somewhat deficient in this. The murderous tendencies of many dogs towards sick or weak sheep did not endear them to me, and although I always had a team good enough for my work, I never became a dog lover or a first class hand with dogs. Nevertheless I had some good canine friends amongst whom 'beardies' rated highly.

Chapter 14

BUILDING A HOMESTEAD— DEATH OF A KAURI

Some four years after taking up my property I became engaged to be married, and it was obvious that I would have to build a proper homestead with rather better conveniences than my 'batch' provided. I had no money, and a substantial bank overdraft, but the farm was starting to produce well, and I decided that, acting on the well-known New Zealand principle of 'do-it-yourself', I could manage. My plan involved felling my own kauri and totara trees, hiring a team of bullocks to pull the logs out, and then a five-horse waggon team to haul the logs to the mill for cutting to my order. This took a lot of arranging, but meant a large saving as against buying sawn timber. I set about sorting out suitable trees; one of these was quite a large green kauri.

Come with me now to see how we tackle this. My 'offsider' is Ralph, a young man who does not know much about bush work, but is learning fast. As we leave the hut for the day's work we each have on our backs a 'pikau'—a sugar-bag rucksack containing tucker for our lunch and 'smoke-o's'. We each have an axe, and Ralph carries a 9-foot two-man crosscut saw newly sharpened and set. I carry a heavy iron-ringed maul and a bag of steel wedges. The early morning sun glistens on the myriad shades of green of the standing bush, and the tui trills his liquid notes, as we set off up a rough bush track. Our objective is a large isolated kauri tree which I expect to provide most of the weatherboards for my future home. We now approach this tree, whose immensity is impressive. The huge column of the trunk, sheathed in scale-like silvery bark, rears without any perceptible taper fifty feet to the first branch, and averages about seven feet in diameter. The tree's large crown of heavy branches with mustard-green foliage spreads out like an outsize umbrella over the lesser trees around.

My heart quails slightly as I contemplate this monster. I have watched experienced bushmen fell similar trees, and understand the technique, but this one for the first time is minc alone, for which I have to take full responsibility. I am uneasily aware of the various things that could go wrong if I mishandle or misjudge the felling of this tree. It would for instance be an unpleasant business if this massive bole tipped back on top of us instead of away from us, its executioners, as it is intended to do. However, adopting a reasonably confident air designed to reassure my mate that I know what I am doing, we get on with it. The first thing is to decide in which direction to drop the tree. The main factors in this decision are—the lean of the tree if any; the weight of the crown—is it heavier on one side than the other? Then there is the surface on which it must fall—is it clear of humps and hollows? The falling bole of a tree weighing many tons, if it hits a mound or straddles a hollow, may break in half or shatter and split much valuable timber. Wind if any, must be considered, for even a moderate breeze can exert quite an effect on the foliage of these large trees, towering as they do above their neighbours. Fortunately in this case the top limbs that appear to be the heaviest are on the side of the tree facing the most level ground, so we decide to fell the tree in that direction.

We commence by 'scarfing' the trunk at its base a foot or two above ground level. This is a V-shaped horizontal cut across the line of the fall, and on that side. The level floor of the cut is made with the crosscut saw to a depth of about two feet. The wood above the cut is then chopped out by both of us, one working on each side. Our axes being sharp, the chips fly in large chunks. As the trunk is rather over 7 feet in diameter where it has to be sawn, I decide to put in a 'side-scarf' to give a better 'run' for the saw, even though this is a 9-foot one, larger than normally used. When this is done, all is ready for the main cut to be put in from the back of the tree. It is important to have this cut level and true, and 2–3 inches higher than the floor of the front scarf. A ring of bark having been chopped out, the saw is carefully started, a tendency of the middle of the long saw to sag being corrected by a couple of wedges driven in to give it temporary support.

Soon the back of the saw disappears in the cut, as with a steady rhythm each of us, one on either side alternately pushes and pulls his end. As the creamy sawdust builds into little piles, the musical ring of the tempered steel blade blends with the bird song of the bush. There is now and then a pause to check that the cut is running true, and, when the saw is well buried in the wood, two of the steel wedges are driven into the cut behind it. They are spaced several feet apart, as a safeguard against the tree sitting back and jamming the saw. As the cut deepens the tree lifts a fraction and lessens the friction on the saw.

As one's body swings somewhat monotonously back and forth and the steel blade sings its 'ring-rang ring-rang' as it bites deeper and deeper into the wood, one's mind has time to contemplate the drama of what is happening. This handsome giant of plant life, which will yield about ten thousand feet of high grade timber, is amongst the oldest living things anywhere on earth. This tree is probably eight or nine hundred years old. Its life is now to be brought to an end by two men, with a thin strip of steel, in a few hours. A tree does not grow for ever, and must ultimately mature and die. I feel a vague sense of regret at having to play a part in hastening the end of such a majestic process. The saw is now well through towards the front scarf, and the tree is lifting perceptibly on the wedges as we drive them further in from time to time. We keep checking that the cut is running parallel to the inside of the front scarf. This is the time when, by leaving a wedge of timber uncut at one side, one can, if necessary, pull the tree to one side or the other when it falls.

There is a crack like a pistol shot, and a tremor in the foliage. I feel a certain tension which I think is shared by my off-sider. This is the critical time. The sawcut is only a few inches from the front scarf, and the small width of timber between the cuts running from one side to the other will form a hinge when the tree tips over. A few more drags with the saw, which we then pull back in the cut. The wedges should do the job now if our estimate of the weight of the crown is not 'off the beam!' 'Bang-bang-bang'—a few healthy wallops alternately on each wedge.

There is a tremor in the trunk, then almost imperceptibly at first, but gradually quickening, the huge column above us tips, appears to sway, and then falls faster and faster as the foliage of the crown swishes through the air. The butt flies off the stump, there is a resounding crash, and with an earth-shaking thud the trunk hits the ground. Thus ends the life of a kauri; and the bird song is stilled.

So we carried on, felling and crosscutting trees into logs of suitable lengths averaging about 16 feet. The nose of each log had to be 'sniped' so that it would slide easily behind the bullocks, and any obstructions or bark was chopped or peeled off to give a smooth surface. I hired a team of twenty-four bullocks and their large Maori driver, Jimmy, to haul the logs the one and a half miles across country to the road where the waggon team could load them. Jimmy was an artist with his team, and knew just how and where to apply the tremendous slow pull of which a bullock team is capable. His two leaders were rangy brindle animals with bright intelligent eyes who responded to his every command. His bullocks were his pets as well as providing his living, and he took great care of them. One weekend I had to put a few cows temporarily in the paddock where his team grazed, and an enraged bullocky lost no time in letting me know that he could not have his bullocks robbed of their tucker like that.

Ultimately he delivered some forty odd logs, large and small, to the roadside. Here a teamster with a fine team of five draught horses and a timber waggon loaded them, one, two, or three at a time, and carted them to the mill from which I later received back the sawn timber cut to the dimensions specified. This was stacked for some months to season properly. All this work was paid for at an agreed rate of so many shillings per hundred superficial feet on the log measurement. I found this timber work very interesting, and got quite a thrill from starting my house literally from the standing tree. This was before the days of the engine driven chain saw—we used two-handed crosscut M tooth saws 6 feet 6 inches to 7 feet in length and I soon learned that a well sharpened and set saw cut a lot quicker than a dull one. In time I became a fairly expert 'saw doctor' with a consequent saving of labour. I even found myself dropping into the jargon of the old hands, who always used to amuse me with their lofty description of a towering tree as 'quite a good stick!'

I had always admired good wood panelling, and determined to have a panelled living-room. I picked out a large twisted rimu tree, that provided enough $12 \times \frac{1}{2}$ inch boards of beautifully grained timber for the purpose. In the spring of the year 1924 I started to build my homestead. I had two permanent hands working for me by then, assisted by a newly engaged carpenter. The routine work of the farm had to be carried on, but we got in every minute we could on the building.

Building a house is a fascinating job, not very difficult really, a task engaging one's creative ability of both hand and brain in a way that makes the undertaking a very satisfying one. Being heavily in debt, with a large bank overdraft on current account, I had to economise all I could, but I did not want to 'spoil the ship for a ha'porth of tar'. In pursuance of this I wrote to an uncle in Australia connected with a wholesale hardware house with a branch in Auckland to ask if he would authorise my obtaining hardware at wholesale prices. I explained that I would need roofing iron, piping, nails, paint, stove, bath, etc. which would run to a substantial amount. He replied that he intended to give me a wedding present, and thought that rather than send something that might be ornamental but useless he would be glad, instead, to foot the bill for the hardware requirements of my house. I have never forgotten this, coming when it did—it was the best present I ever had, and enabled me to finish the house without skimping.

In April of 1925, after six months' work and almost five years to the day since I had taken over the property, the house was completed. It was built throughout of heart timber, mostly kauri and totara, and stands today forty-three years later as good as the day it was built. When it was finished I got a kick out of just looking at it standing there amongst the blackberry and rubbish that would have to be cleared to make a garden.

I have dealt with the building of my homestead at some length, because it is an example of the versatility that a farmer in the backblocks has to acquire to be successful. Unless he has plenty of capital, which is rarely the case, he cannot afford to employ many skilled tradesmen such as carpenters and plumbers who have to be paid high hourly rates, a penalty rate for rural work, travelling time and transport costs. He therefore has to cultivate the 'do-it-yourself' idea in all sorts of directions.

We ultimately built a full range of farm buildings, including two more houses for employees on the same lines as the homestead, sometimes doing the whole job ourselves, but occasionally employing a carpenter to help us out. Similarly the timber, nearly all kauri and totara, used in these buildings came off the property. This was normal practice in most northern districts in the early establishment of farms. Now we have practically exhausted the valuable native timber except for a few specimen trees that I have reserved, including some green kauris in a very inaccessible spot. The further development of the helicopter will probably result in their being lifted out in slings in years to come, though I am glad to say my son is as jealous a guardian of a good tree as I am. Into our latest farm building, additional quarters for shearers, went some of the first of our home-grown exotic timber, pine and eucalypt, from plantations I put in thirty-five years ago.

Chapter 15

FAMILY AND FARM

Then commenced my married life in which I was to experience the loyal help and encouragement of one who was no stranger to farming. My wife had her own horse and dog, and delighted in helping me with the stock work. These were the days before electricity had reached the backblocks. We used candles and kerosene lamps, and cooking was done on a wood-burning stove, with a 'Mrs. Potts' charcoal iron for ironing the clothes. We were young and 'hard up' but had good health and plenty of energy which we used in creating a garden out of the rough stuff surrounding our new home. What a great time in life it is when one is young and happily married, made even more interesting by the arrival in due time of a family, in our case two girls and a boy.

This would be a good place to pay a tribute to the splendid farm housewives of New Zealand. We men get the credit for enterprise in developing land, but behind nearly every man is the capable wife with great capacity and able to turn her hand to almost anything, besides having to put up with us in moods of both exaltation and depression. The girl reared on the farm has an advantage—her education and ability in a lot of things is far wider than that of her town sister. On the other hand many town girls who marry farmers learn quickly, and develop all that is best in them in the wondrous and varied environment of country life.

In the early days they often had much to put up with. Roads were poor, and horses provided the only means of transport. The farm wife baked her own bread as a matter of course, and churned her own butter. Often there was no telephone, and we forget that wireless is a comparatively recent innovation. Schooling for the children was difficult, usually involving helping the kids to catch ponies after giving them breakfast and cutting their lunches. The men on the farm were often working some distance away, and, busy as she might be, the housewife had a lonely day. Our first two youngsters double-banked to school on a shetland pony called

Dandy, who was a dear little chap but would stand no nonsense. One day they were late home from school and obviously a bit upset. With a concerned air our small daughter aged seven explained that 'Dandy played up—he'd even made marks on the road!'

One by one modern amenities have made farm life easier. Roads have improved—no longer do we travel in a gig or buggy at walking pace to the sound of 'glug glug' as the poor old horse plods through twelve inches of glutinous mud. We have metalled roads and motor transport. The ability to drive a car has been a great thing for farm wives in giving them more freedom and reducing isolation. Electricity, now reticulated into all but the most remote spots, has lightened the tasks of cooking, heating, and cleaning. The advent of electric power for the first time in a country district was a memorable occasion. After years of kerosene lamps and candles with their smell and indifferent illumination, it seemed a new world at night. I had been on the farm for eighteen years before we received this boon that is accepted by the town dweller as a matter of course. I well remember the evening when we were told the power was 'on' for the first time. It was marred by only one slight contretemps. We pressed the switch in the kitchen, the light came on in the living-room. A slight mistake in the wiring!

The radio receiver giving out music, talks or what-have-you, is good company for the housewife when the men are working out at the back of the farm. Then there is the party telephone, which if all else fails to disperse some frustration, is well known as a splendid means of emotional release if you have one or two sympathetic neighbours. Now there is television. As to whether this is to be claimed as an advantage I am unsure.

A lot of interest centred round the radio and the party telephone on a remote farm. These were the points of contact with the outer world, and both played a large part in countering the sense of loneliness and isolation that used to be one of the trials of life in the backblocks.

Our first radio was a three valve battery set, which I brought home anticipating all the entertainment we were going to have. I duly installed it, turned the switch and—'Phut'—blew all the valves. When we got it going the normal reception was often drowned out by the somewhat raucous voice of a local gossip on the party telephone coming through the receiver. This embarrassing and unwanted reception was traced to the too close proximity of the earth wires of the radio and telephone.

Anyway the net effect of all this today is that the farmer and his wife have the best of both worlds if only they realise it. They have nearly all the worthwhile amenities that the town dweller has, without having to suffer the many disadvantages of town life. We don't have to put up with the neighbour peering into our privacy or despoiling our weekend hours with the sound and smell of his too noisy motor mower. We don't have to take the noise and stink of petrol and diesel fumes that are a concomitant of town traffic. We do not have to drink chlorinated water, and we can even make a cup of really good tea with rain water. In most places we have a daily goods and mail service, even though the delivery may not be to the door. And there is not an excessive number of people crowding us at every turn as in packed buses needed to reach a job miles from home.

Finally we are able to live in the unpolluted clean air of the country, with grass, trees and stock, around us with all the marvellous interest in things to do and see that such a life provides. Why don't we make more of all this, particularly where young people are concerned? Every year, presumably under the guidance of vocational officers, we see thousands of young people leave school and head for blind alley or parasitical occupations while only a small minority opt for a country life.

There is something wrong here—more of it later in the chapter on Farming as a career. Farm children may be slightly behind town children on the academic side of education, due to smaller schools or sometimes indifferent teachers, but the environment around them gives them a great range of interests. They have their own farm pets, with all the natural life in plants and animals which is an education in itself denied the town child. When my children were young, timber was still being worked by bullock teams. My small son did not have the usual urge to be an engine driver—what he wanted to be was a 'bullocky' with his own team. To this end he reared two Hereford steer calves whom he trained to pull a sledge with a proper yoke and bows which I made him.

A fresh lot of larger gear had to be made as the 'bullocks' grew. Ultimately he had two big lumps of steers, 'Goggles' and 'Mousey', who could pull a fair load on a sledge. Complete with a long whip and a hardcase old straw hat, the 'bullocky' looked the part. Like most pets these animals did not get enough work, and tended to get a bit uppish. There would be the sound of galloping hooves and

Aerial topdressing on the central plateau of the North Island—Mount Ruapehu the skiing resort in background.

The loader drops its three-quarter ton load of phosphate into aircraft hopper prior to take-off. This operation takes seconds only.

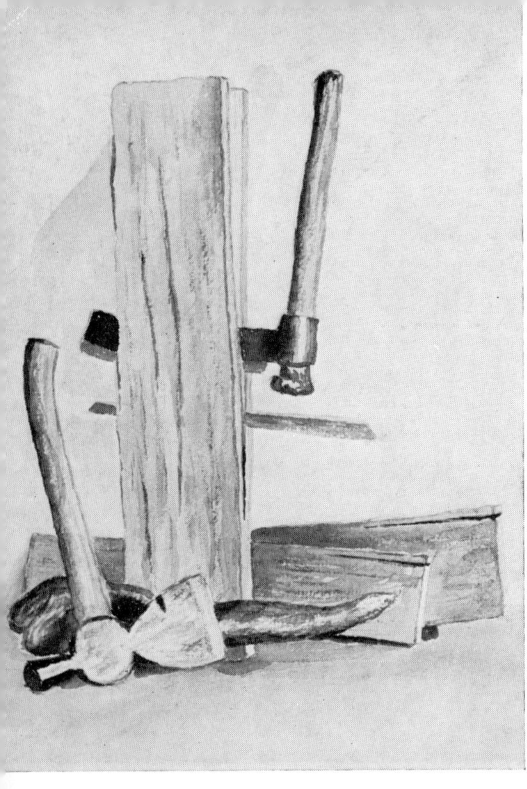

Wash drawing by author showing use of the 'frow' or paling knife to split shingles. Shingling hatchet and kauri knot used as a mallet in foreground.

Before there was a road or a woolshed. These sheep are fording the Hikurangi river on the way home after being shorn ten miles away.

shrill cries of rage, followed by a loud jangling crash, as the team and sledge collided at high speed with the fence. The team had bolted again. Another accident with the team resulted in some broken gear, and a tearful small boy announced to his horrified mother that 'me bows are buggered'. Came the time when the steers were too big to keep around, and they had to go to the sale-yards. In a regretful farewell their master said he hoped they would be bought by 'another bullocky'.

Small children can be both very trying and a lot of fun. Following an altercation just outside the back door one day there was a brief silence, and then an apology from our conscientious small daughter to her younger brother. It was 'Sorry I spat in your eye—I only meant to spit in your face'.

In the first few years of married life when we had to depend very much on horses, my wife either rode or got about in a gig with 'Spick' the pony. This outfit was given her by an old friend of her family. Then we were able to buy our first car, a second-hand light English one, because it was cheap. It was not until 1936, when wool prices took a much overdue turn upward, that we were able to revel in the luxury of our first new car. We used to sit in it just to sniff the opulent smell of leather (yes . . . real leather) upholstery and new paint—that 'new car smell'. What a thrill!

Just as a car made all the difference to us in the late 1920's so had the coming of the railway to Whangarei a few years earlier enlivened the town and district. More people were attracted to the North, and the influx of manpower, 'know-how' and capital speeded up development.

The railway was of difficult construction involving many tunnels. Maoris were less sophisticated then than now, and the story is told of Hori, straight from the pa, who had never travelled by rail before, but decided to see the world by taking a train to Auckland. He was thoroughly enjoying himself, when, without warning, the train dashed into the first tunnel. This was something outside his experience; his speech faltered and he looked wildly around as one seeking a way of escape. At the height of his agitation the train shot out into daylight. With the resilience of his kind, Hori beamed upon his fellow passengers, and remarked with relief—'by korry . . . tomorrow, eh?'

Chapter 16

HOW THE FARMER SEES IT

Our attitude to life is very much conditioned by our environment, and this is particularly true of the farmer. He is 'self-employed' as the present ridiculous term has it, administering, working, and developing his own little empire. He thus becomes very much of an individualist. His place in the economy of New Zealand is a paradoxical one. He produces some 93 per cent of the exports that provide our foreign exchange, which buys all our imports, and thus plays a very important part in the standard of living of the whole country. On the other hand he is in a minority electorally, and has to submit to policies determined by a government, of whatever party, that relies for its support on the urban vote of the townspeople and non-farmer sections. These groups have considerable protection of their earning power by way of protective tariffs for manufacturing industries, price-fixing rings, and the setting of minimum wage levels by an arbitration court whose decisions have the force of law.

The New Zealand farmer, who, incidentally is the highest *per capita* producer of primary produce of any country, has to sell in the markets of the world. When these take a down-turn his gross returns fall, but he has to continue buying all his requirements in labour and materials in a highly protected local market. His costs thus tend to be rigid, and he finds himself ground between the upper millstone of lower produce prices and the nether millstone of high costs. The net result of this is that his income fluctuates wildly, usually tending to fall at the most inconvenient times. When is added to this the vagaries of climate such as droughts and floods, the effects of stock diseases or other unwelcome visitations, something of what he has to contend with will be seen. It is due to those factors that the 'projections' of the economists on farming income so often fall flat on their face.

Farming land having been driven up to a high value by the continuing devaluation of the currency, or inflation, which seems to be the inevitable policy of various governments who wish to maintain themselves in power, the farmer has a further problem. This stems from the fact that he has to finance his operations on borrowed money. He thus invariably has a fixed commitment as a mortgage on his land, and in addition commonly finances his seasonal operations by borrowing on bank overdraft. Interest on these borrowings represent a more or less fixed annual charge, which has to be met, whatever prices he receives for his stock, wool and dairy produce.

All these things colour his outlook and his attitude to his fellow men. When prices fall through no fault of his own, and his income drops drastically, he often feels that he has been the victim of some economic confidence trick. At such times he tends to be envious of others whose wages, salaries and profits continue unchecked. He also notes that whereas he is paid by results, that is on the quantity and quality of his output, the wages and salary man is paid on time worked, which sometimes has little relation to his production. I feel these matters should be mentioned because too often the farmer is thought to be an unreasonable fellow by comfortably situated people who do not understand his circumstances.

In recent years farmers generally have sought to keep their heads above the waters of financial difficulties by increasing their production, and what is shown by the official figures is very much to their credit. This has involved the working of long hours in the application of modern techniques, and the investment of any income available after paying his taxes and living expenses in development work of one kind or another. He is often helped by his family, and wives have played a noble part in this. Especially on 'one man' farms, not only do they run a house, provide meals and look after a family, but take an active part in assisting their husbands with their work. They can usually drive a tractor or a truck, ride a horse and help with the stock. My own wife did all these things and had her own sheep dog, and without her help I would have been sunk.

Looking back on some fifty years of farm life, I found the physical work and all it involved troubled me little. All the worry and anxiety was associated with the effort to meet financial obligations in times of uncertain and fluctuating prices. The farmer deserves any help he can be given to lighten this worry. Loans for development work on a longish term basis which cannot be called up suddenly are now being provided. I found the catch about overdraft finance was that the advance was at call, and if the banks panicked, as they sometimes did, or were involved in what is euphemistically called today a 'credit squeeze' one could be asked to reduce or repay what one owed at short notice. I had more than one such unpleasant experience. Stable prices in a fluctuating world market are impossible to arrange, but the fixed seasonal price for dairy produce, and the floor price for wool, now operative in New Zealand, give some help that was not available to earlier generations.

In our age of division of labour and the 'organisation man' every section of the community of any economic consequence has its own union, guild, association, or what have you, to preserve and enhance its interests. Governments take little notice of individuals, but they do take notice of well organised groups, especially where many votes are concerned. Such bodies can and do play a part in moulding national policy according to the amount of political pressure they can bring to bear.

I early formed the opinion that the farmer was losing out here because his case was not adequately heard or presented. Despite my other preoccupations I became involved in this during the early stages of breaking in my bush farm. At first a local leader in the principal farmer's organisation of the day, as the years went on I graduated to provincial level and then to the national body which met in the capital city. This was an educative experience. As noted earlier, the farmer is, by virtue of his calling, very much of an individualist. He has plenty on his mind, often with his nose well down to the grindstone, and he does not take kindly to organisation. This meant that while some would not even join their own set-up, though quite ready to profit from any results it obtained, others had to do a good deal more than their share. The 'willing horse' was worked pretty hard, and though I could ill afford the time for these activities, the great pleasure of it was the association over many years with a splendid body of men all working unselfishly for the betterment of their industry.

One who is elected to represent his fellow farmers has to be prepared for all sorts of criticism. There were those who said to me 'You are a mug to be giving your time and energy to this sort of thing when your farm is far from developed'. Then there were those of ungenerous mind who said 'Well of course you must be getting something out of it—I bet you do yourself pretty well on these trips to Auckland and Wellington.' It was good self-discipline to try and maintain an unruffled appearance whilst often inwardly fuming at such jibes. In pressing our views on the Government, and trying to get the farmers' problems adequately dealt with, we often seemed to be butting our heads against a brick wall. In retrospect however, looking back over many years, I can see that our determined and repeated representations to various governments have had worth while effects.

In the last issue the farmer is on the box seat in New Zealand. Our whole topheavy economic structure is based on his production, and other sections must concede him at least reasonable conditions if he is to continue to come up with the remarkable output of recent years. To do otherwise is to kill the goose that lays the golden eggs.

An interesting facet of representing the farmers' position to the government were the many deputations we had to various ministers. During the long reign of a labour government from 1935 to 1949 I was involved in many of these. We often had to see Mr. Walter (later Sir Walter) Nash who was minister of finance and marketing. There he would sit at his desk, behind mounds of files, with an ironic smile on his face as we put our case. He knew perfectly well that, politically, we were in the opposite Camp, and one could almost divine his thoughts behind that inscrutable smile. They were undoubtedly something of the order of 'You boys would put me out tomorrow if you could, but all the same you've got to come to me if you want things done!' To his credit I am bound to say that he always gave us a fair spin, he was always well informed on the matter in hand, and any commitment he made was scrupulously observed. The knighthood bestowed on him shortly before his death was well earned.

With others I took part in originating farmers' co-operative companies to market both meat and wool. Most of these now have a substantial throughput and receive wide support. The advantages of co-operative handling of our produce are so obvious, though demanding some self-denial in the early stages, that I am always amazed that farmers take so much persuading into it. As one of a generation now fading from the scene, I reflect that we have left the younger men good organisations to carry on and develop further, though I doubt if they appreciate the difficulties of getting them going in the first place.

Chapter 17

LOCAL COLOUR

No record of this sort would be complete without reference to the people amongst whom one has lived and worked. Each district has its own characteristics, and that in which I have found myself has a splendid community spirit. Anyone in trouble does not look for help in vain, and all deserving causes can count on support. The early settlers were mostly ex-bushmen, who, when the big stands of kauri were worked out, had been able to buy land cheaply. Rivers were unbridged and the unmetalled clay roads were few and far between. Many of the early families were of Nova Scotian origin, and there were some Belgians. It is related of Angus and Dan Finlayson of Waipu, that they bought their land in bush in 1882 for 10s. an acre. They had no money for improvements, so they went off to work in the kauri bush at Mercury Bay to earn enough to clear and grass a small area. They were away for two years. This is typical of early settlement of these parts, once an almost unbroken area of forest. The first buildings were usually constructed from split palings of kauri or totara, which stood for many years. These were followed later by houses built entirely from pit sawn timber. As a pioneering feat of self-reliance the following, written by Mr. Roddie Finlayson in a short history of the district, takes some beating—'He then felled quite a large kauri tree, *20 feet in girth*, single-handed. From this tree he built a slab hut.'

The centre of community life is the local hall. This was built in 1897 by the free labour of the settlers, the timber being all pit sawn from the log, and hand-dressed. There it stands today, sound as a bell, a splendid bequest from the early pioneers to those who have followed them.

In those times there was a big trade in kauri gum, dug up from the 'gumland' which had once been kauri forest. When I arrived in 1920 this was still going on, and there was a considerable trade in it through the store-cum-post-office at Moengawahine, at the end of the clay road. A big event was the twice weekly arrival of the mail and stores at midday. This was carried by a buggy and pair from a livery stable in Whangarei. As the owner reckoned this was a good opportunity to steady down some of his newly broken-in horses, passengers often had an interesting journey.

Round the store would be clustered a mixed group. There would be Pakehas like myself to get their mail, but the majority would be Maoris from their settlement a few miles further on. There would be an assortment of horses, many with split sacks across the saddle, and the odd pack-horse, from which were unloaded the bags of carefully scraped gum. These would be tipped out on the floor, to be closely assessed for grade, and then weighed, by the Belgian that ran the emporium. The customers having received their credits for gum supplied, took them out in groceries which were loaded on to the poor patient old nags.

There was lively chatter and repartee, and much laughter, especially when one Maori housewife was detected getting away with more than her share of groceries! News and gossip was freely exchanged, and this was a good time to find out what had happened to Hori last week when he didn't turn up as promised. 'Him. Oh he go to the tangi* at Titoki—I not see him since'. 'Where's Eru? Him gone to help his Uncle—him very old man—I think he be away long time.'

So it goes on, till, the fount of news and entertainment being exhausted, they gradually drift away to plod slowly home through miles of the glutinous mud clay roads. It is my guess that the Maoris got just as much fun out of life then as they do now, when a daily bus on a metalled road runs through their settlement.

An interesting commentary on prices in the early days of settlement, in 1889 and 1890, is provided by extracts from the diary of John Pullin, whose descendants still farm in the district, as given below:

Prices Of Farm Products And Wages At The End Of Last Century.

In John Pullin's diary, prices of wool and sheep in 1889 and 1890 were as under:

Ewes: 5s. to 7s.

Lambs: 5s.

* Tangi—funeral feast.

Wool shorn from 50 ewes totalled 158 pounds at 7⅛d. a pound brought him £10 8s. 11d.

In 1892—86 ewes and 8 rams produced two bales of wool at 6¼d. a pound returned him £18 13s. 3d. for the 758 pounds.

Girls that he employed to help in the house in 1886 were paid 9s. a week. In 1887 Maggie Morris for twelve weeks less three days received £5 in wages.

Bush felling cost £1 per acre.

For the fencing contractor the farmer cleared the line to a width of 12 feet. All fencing material had to be supplied and put on the line. The fence erected by the contractor cost the farmer 8s. to 10s. per chain in wages.

We had our share of 'characters' not the least of whom was 'Old John' who had been on one of the older established farms most of his life. He maintained that he had been 'sold with the farm' twice. He was one of the old type 'station hands', long past his prime, but kept on to do light work. He did some gardening, but it was not his strong point. The farmer's wife having planted a lot of seedlings, she was greeted a day or two later with the announcement 'I weeded all them little weeds out for you Mrs. — and that was the end of them. He always referred to nasturtiums as 'extortions'.

A friend of his having been taken to hospital, he reported that 'he had to have blood confusions'. He lived in a small hut of his own, and, being keen on racing, derived much pleasure from the racing broadcasts made by Gordon Hutter. As he expressed it 'no sooner was one horse out of his mouth than there was another one in it!' Trespassers having caused trouble by leaving a gate open, John put up a notice which read 'Nobody must not leave this gate open'.

One summer he was engaged in discing a paddock in clouds of dust, and his boss in passing remarked 'You'll have to have a bath tonight John' which drew the reply 'Oh it's all right, I've got me pants tied at the bottom!'

Cooking for some of the men he was criticised for over doing the salt in the stew. Next night there was stew again—without salt —and he announced. 'Them as wants salt can put salt in, and them as don't want salt can have no salt!'

Bees on the rampage could cause incidents that were humorous to those that did not get stung. One of my neighbours fancied himself as an amateur apiarist with half a dozen hives. One day I called in to see him as he was busy with his bees. I kept a respectful distance, but he called out 'It's all right, Briscoe, they won't hurt you, it's just a matter of knowing how to do it and taking them quietly'. I edged forward another step or two, and at that moment several bees stung him simultaneously. I could not restrain my laughter, but it was not appreciated.

Bees got into the wall of our men's quarters, between the outside weatherboards and the lining, and stung one young fellow on the face. He was evidently allergic to the poison, his whole face swelled up and his eyes closed, giving my wife and myself a good deal of worry till he got over it. We decided we would have to take a bit of the wall down and root the bees out. At this stage we called in Bill to help. His father had been a cowboy who had come to New Zealand from the state of Utah in connection with the establishment of the Mormon church in the local Maori community, and his mother was a Maori of good family. He was a splendid type, over 6 feet and weighing about 16 stone, good natured, and on for anything.

He was particularly good with stock as might have been expected from his cowboy ancestry. When we were working young cattle in the yards, and a beast had to be held, Bill was in his element. He wrapped himself around a yearling beast with no apparent exertion, and there was no further argument so far as it was concerned. 'Bees? No trouble at all boss—we soon fix them'. Upon which Bill proceeded to wrench several weatherboards off the building while the rest of us remained at a distance. There was a wild flurry of bees, several of which stung him. With a large grin he brushed them off imperturbably as though they had been flies, and proceeded to clean out the wall cavity where they had been such a nuisance. He must have been stung many times, but it didn't seem to worry him.

For many years, until I could afford to put in a pump, we relied entirely on rainwater from the roof of the homestead. In dry spells the tanks ran out and we had to cart water from the creek. This Bill did with a technique of his own. Sitting on the top of two forty-four-gallon drums chained to a wheeled sledge drawn by two horses, he would drive to the creek, remove the upper bungs in the drums, and then drive into a deep hole in the creek. Under his weight the whole contraption sank quickly, the drum filled with water, and Bill, soaked to the waist, but grinning hugely, drove out with eighty gallons of water.

Barney, an Irishman who worked for us for some time, was keen on bees, so I built him a proper hive. As I was completing it outside the workshop, while Barney watched, a stray bee landed on the entrance step. 'Ah,' said Barney, 'them bees they knows.' They certainly knew how to collect the nectar from the flowers of the eucalypts, which produced honey of the most delicious flavour. Barney had been involved in not a few brawls. His recipe for victory was—'take yer teeth out first—then pull his hat over his eyes, and slog him before he can see yer'.

Another employee had a sharp-tongued wife, but did not take her caustic comments too seriously. Once turning up late for work, he announced 'the gestapo's been on the rampage again'. Cattle that took to the bush became very wild and difficult to handle if you could get them out. Tommy, a well-known personality, who had his own team of bullocks, was an expert both at recovering them from the bush and 'civilising' them. He had dogs that would grab a beast by the nose and pull it down, and to protect himself against a beast that charged, as some did, he carried what he called his 'dream tablet' a large heavy 'waddie' which he did not hesitate to use on the head of a dangerous beast.

One day riding down our newly formed road a Maori approached from the other direction at full gallop. He slowed up enough to tell me he was after some rata vine, the sap from which is a Maori remedy for stanching bleeding. 'Hauraki he blow his hand off blasting fish with gelignite—you go and see him'. I got a move on and found Hauraki with most of his hand gone and only the shredded tendons of his fingers showing—a gruesome sight. We gave him what first aid we could and got him off to hospital. A high price to pay for carelessness with detonators and explosive. The Maoris have a whole range of herbal remedies based on our natural flora. They held the delicious toheroa in high esteem for restoring an invalid's health after an illness. Packed with proteins and vitamins as this shellfish is, they could not go wrong.

I had a hard case stock pony called Clarence who had a very hard mouth. When drafting cattle, galloping after a breaking beast, he was very hard to pull up, and the men used to delight in telling my wife stories of the boss 'going like the wind over the ridge in the direction of home'.

Lolloping home at an easy canter on 'Clarrie' one day, my mind far away, I rounded a blind bend in the road to encounter my neighbour driving a sledge. Clarrie propped violently on his fore-legs, and shot me off into a blackberry bush on the side of the road. I didn't hear the end of that for a while.

It was many years before all the roads were metalled, and there was constant pressure on the county council, who had to do the best they could with meagre financial resources. One settler who lived at the end of a long clay road wrote a despairing letter to the local paper in the autumn. In it he said that since there appeared little chance of his road being metalled before the winter, they would just have to do the usual thing and 'order a ton of flour and two dozen mousetraps' before his road became virtually impassable.

This kind of life produced strong characters of both sexes, full of resource and often able to extract a wry humour from their difficulties. One such was a lady who could 'dish it out' with a caustic tongue. She took to a neighbour of mine one day to such effect that he swallowed his cigarette butt. As he spent the next few minutes choking, his replies were ineffective and she had the best of the argument.

Dave was an elderly farmer whose sight necessitated the use of glasses for most things. Mutu was a genial Maori shepherd with an over developed sense of humour—to him life was a huge joke. In the course of yarding a big mob of ewes and lambs for drafting at shearing time, Dave lost his glasses, insecurely carried in a hip pocket. A frantic search failed to reveal them, but Dave thought he could draft the lambs off the ewes without their aid.

As he forced the sheep up to the drafting race, Mutu recovered the glasses, undamaged, from the dust of the yards. Watching his chance, as the sheep approached the race, he carefully adjusted them across the nose of a steady old ewe which he shepherded forward to meet the outraged gaze of his boss at the drafting gates. Dave being deficient in a sense of humour 'blew his top' and Mutu narrowly escaped being sacked on the spot. Everyone else enjoyed the joke.

Chapter 18

DISASTER—
THE SLUMP OF THE 30's

The great slump of the early 1930's was a tragic experience for many countries, but for none more than those exporting primary products. New Zealand's whole economic welfare rested then, as it still does today, on the sale overseas of butter, meat, wool, cheese, hides and skins, fruit and other items. The prices of these fell to very low levels, relatively much lower than those of other goods and services. The result was that farmers could not meet their interest payments on mortgages, and had to stop buying all sorts of things normally used but for which now they could not pay. This threw out of employment thousands of men in the import, manufacturing, and distributive trades, until the whole economy of the country was in a desperate state. Those who were able to retain State or other salaried jobs, even with some cuts, were relatively the least affected, because in terms of real wages, what money would buy, they were comparatively well off.

In these circumstances our much vaunted system of wage fixing by an arbitration court did not show up very well, because of the rigidity of the legislation. Many employers could have kept men on at reduced wages they would have been glad to accept, but this they were not allowed to do. They had to say, to many men: 'The law says I must pay you a minimum of so much in wages. You know as well as I do that in the prevailing circumstances I am unable to do this, so I am forced to let you go.' When the slump hit farming, I had been farming about eleven years. I had increased the carrying capacity of my farm and its output six fold, but technically I found I was now bankrupt in spite of all this, because I could not meet my financial commitments. Whereas before I had reckoned to meet my interest payments on borrowed money, by the sale of twenty-five bales of wool, I now found that it would take sixty to seventy bales of wool, which I had no hope of producing. My situation was typical of that of thousands of farmers.

The psychological effect was shattering, particularly on returned soldier farmers like myself. We felt we had done the right thing by our country in war, and subsequently in the years of peace by increasing the country's exports. Now we were all at the mercy of those to whom we had financial obligations—the Government, banks, insurance companies and other lending institutions, and private individuals. This was through no fault of our own, but because of some failure in the money system. It was too much for a friend of mine—he went out into the scrub and shot himself, leaving a widow and family. The general effect was to make us feel that if this could happen all effort was futile.

Being of an enquiring turn of mind I wanted to know *why* this had happened, feeling that if we knew the cause something could be done to mitigate the trouble. It was clear there was no failure in the system of physical production—enough food and manufactured goods were being produced for the needs of mankind, but people were not having their needs supplied because the means of distribution, the money mechanism, had broken down. Why was this? Research showed that it was clearly the result, in the main, of a panic by banks throughout the world, which had been triggered off by the failure of a prominent bank in Austria, followed by many more. It was not then so fully realised, as it is today, that the main component in our financial set-up is not the cash and notes that we handle in minor daily transactions, and which form but a very small proportion of money in use, but bank credit.

This, the main body of our money mechanism, was controlled by the banks, and operated for profit. At times of expansion of bank credit, the banks created credit, or new money, by the granting of new overdrafts over and above their lending existing deposits. When credit was restricted, overdrafts were called in, and a proportion of such credit, or money, was destroyed. This issue and recall of credit for commercial exploitation is hard to justify. It surely should be the prerogative of the national Reserve Bank, acting in the interests of the community, to control and regulate both the issue and recall of credit. Today the Reserve Bank in New Zealand exercises a partial control through the 'reserve ratio' system, which puts a brake on excessive credit creation, but this does not go far enough. As a result the banks do not have quite so much freedom, and a very good thing too.

The slump was precipitated by banks throughout the world concluding, at more or less the same time, to call in and cancel overdrafts to customers on a large scale, for reasons that I need not go into here, but that seemed good to them at the time. This paralysed activities of all sorts, and led to price falls of commodities because the usual buyers no longer had bank credit to buy with. One authoritative estimate was that the banks had called up and cancelled lendings totalling £25,000 million, or in other words had destroyed this amount of money.

The currency and banking system of each country is its own affair, and it should be the responsibility of the Government to see that it is operated in the public interest. It was the misfortune of New Zealand that when the slump hit New Zealand with its devastating effects, we had a government that politically appeared to be under the domination of the big business money lending interests and that had amongst its leaders no outstanding men with imagination and the strength of purpose to initiate unusual action to meet an unprecedented situation. The so-called 'experts' in the economic and financial spheres all came up with different diagnoses and remedies, nearly all of them lamentably wide of the mark. The Government instituted various relief schemes at very low wages, and stay orders of different kinds to check undue exploitation by lenders of money of those who could not now meet their obligations.

But it would not tackle the one thing that could have given some real relief—the issuing of special credits for specific purposes, such as public works, or subsidies on export prices by the Reserve Bank. The politicians were assured this would never do—it was inflationary! This at a time of acute deflation, when the injection of new bank credit was the one thing needed! I was amongst those who publicly, as a farming leader, advocated this sort of action, but we were howled down as 'monetary cranks'. Today, with the retrospective experience of that time as a guide, I think it is generally accepted that governmental action associated with the tightening or liberalising of bank credit is a necessity in guarding against booms or slumps, but those of us who advocated this in the 30's were very much in the minority.

Over the three or four years for which the slump endured there were many tragic happenings. Men who deserved well of their country lost farms and businesses into which had gone years of effort, and some became embittered by the injustice of it. Lending institutions came into possession of surrendered properties of all kinds at a fraction of their true value, out of which they made millions when values were restored in later years. Public criticism of a government unwilling to take some of the strong measures needed mounted steadily. A number of returned soldier settlers in my area were forced to abandon their farms with all the hopes and years of toil that had gone into them.

After carrying on for some time with postponements of interest and other stop-gap measures, I was forced to contemplate the same action, since there seemed no point in going ahead living on borrowed money which I was unlikely to be able to repay. At this nadir of my misfortunes I was offered a job as a carpenter by the head of a building firm in Auckland, and I realised I had to make a decision. I had been fortunate in that my principal creditor, who had sold me the property and to whom I owed a large sum on second mortgage, was an elderly practical farmer who had been considerate, and understood that it was not my fault I was in arrears. I now approached him, saying that though I did not want to take another job, I felt I could not carry on the farm unless he could make such concessions in my indebtedness to him as would give me some chance, when the hoped-for rise in prices took place, to make good. He proved to be a wise and generous man, and met me fairly. In subsequent years he had no cause to regret this, as I was able further to develop the farm and his mortgage became a gilt-edged investment. We were good friends until his death at an advanced age.

I have recorded these experiences of the slump in the 1930's because it left a scar on the minds of thousands of my generation. A philosopher has said that 'history shows that we do not learn from history' but I think the people of New Zealand did learn something from this experience. They at least learnt how important to public welfare is the operation of the monetary system that governs our economic life, and this has resulted in certain curbs on the activities of private banking. It has also produced a political party that would go much farther in this direction if elected to power. Without endorsing this or any other policy I think it is fair to say that in a highly literate electorate there is now a much better understanding of the necessity of a monetary mechanism that will function in the public interest than there ever was before. This makes it unlikely that we will in the future have to endure a slump of the severity of the 1930's, because public opinion is sufficiently well informed to insist on sensible practical measures.

In turning out some old papers I came across an account sales of 942 pounds of wool sold in Auckland on 13th of April 1932. It is of some interest to reproduce this here as an instance of the hopeless position of the sheep farmer at that time:

					total		
Lot	*bales*	*description*	*lbs. weight*	*price per lb.*	£	s.	d.
302	1	lambs	418	1d.	1	14	10
320	1	crutchings	465	½d.		19	5
354	4 bags	black	59	3d.		14	9
			942		3	9	0
Less total charges (receiving, weighing commission etc.)					1	5	10
				Net proceeds	2	3	2

It was this kind of experience that led us to develop the present 'floor price' plan to give some protection to the grower. For three years on end my annual out-put of forty odd bales of wool, eight hundred sheep and one hundred head of cattle were sold and the total proceeds paid to others, while I had to borrow money to live on. This was the common experience.

One effect of the slump was a great increase in 'do it yourself' activity at which the average New Zealander usually shapes pretty well. Home brewing became popular and one took some risks in accepting hospitality from neighbours who fancied themselves at it. The brew that was a bit heavy could lead to alarms at night. There would be a dull report from the wash-house—'dammit there's another bottle gone'. By and large we got a lot of fun out of it.

Hides became practically valueless, but leather, a manufactured product, was too dear for us to buy. I could not bear to see these good skins going to waste, so I became interested in tanning. Two or three wooden barrels were easily obtained, and wattle and tanekaha tree bark gave me the material for tanning liquor. A tin of quicklime for plumping and dehairing the skins, and I was ready to go. Over a period of months I tanned cowhides, which took about four months in the tan, yearling hides and goatskins. I am still using some of this leather today, thirty years later. A present that was very popular with my friends was a pair of reins from a cowhide tanned in tanekaha bark, which makes a beautiful leather. My horse still wears a pair of these reins.

The bole of a large kauri in a newly-burnt clearing. My horse gives an indication of its size.

The butt log of a dry kauri which had a girth of twenty-five feet.

In the back country of Canterbury.

Bringing sheep into the woolshed.

In the farming world I find that we older people, when we tend to be cautious about future financial commitments, are told we are too 'slump conscious'. The implication of course is that 'it can't happen here' again. It is well, none the less, to have some reservations about this. Our economic dependence on a few main items of primary produce is so great that we could in certain circumstances take quite a knock. We are already confronted with difficulties in selling our steadily increasing production of foodstuffs, which seems a paradox in a world with so many people on a bare subsistence level. Our wool, skins, and hides, have to fight an ever-increasing battle against synthetic products to retain their old traditional markets.

So far we have been able to hold our own, but governmental policies in the consuming countries restricting entry of our products, or levying them with heavy imposts, as is provided for under the present treaty governing the common market, could affect our prices disastrously. Britain has already put on a quota limiting the amount of butter she will take from us, and by manipulation of consignments dumped by other countries has brought the price of butter down. Meat is also under the threat of a 'managed' market. In these circumstances we would be wise to keep our ears pricked, try to hold our costs and keep our financial commitments within sane limits.

Chapter 19

HAND TOPDRESSING— 'WE STAGGERED ROUND THE HILLS'

Signs of better times began to appear during 1935 as various remedial measures in the international sphere began to take effect. A Labour government was returned at the end of that year. New policies were put into action, and as the markets for our main products brightened with better prices, there was a gradual recovery. I have described how the grass on my hill farm, like thousands of others, was established by surface sowing on the ashes of the burnt forest trees. So fertilised, these pastoral lands were rich in plant food, the grass thrived and could carry a lot of stock. As time went on, and stock sold off the land removed more and more of its mineral content, fertility declined, and with it the carrying capacity. The years of the slump accentuated this. In the farmer's impoverished state proper up-keep was often impossible. Many properties were abandoned, and the hard-won pastures allowed to revert to fern and rubbish uncontrolled by the stock that had been sold to meet debts. Mortgagees who had taken over farms were often unable to manage them properly. The net result of all this was widespread deterioration and reversion over millions of acres of hill country.

As I had been lucky enough to keep my property and stock, and had worked hard at keeping fences in order, the land suffered little surface reversion, but I realised that a steady fall in fertility was going on. The obvious answer was to try to topdress a proportion of the farm with phosphatic fertiliser, hoping that there would be a certain transference of fertility to the more inaccessible areas by stock when they were moved out. This was a formidable physical problem which I rather dreaded. The only mechanical aid available at that time was the bulldozer, with which we could make roads some distance out on the farm to the areas to be topdressed. We could make dumps of 10 or 15 tons of fertiliser at the ends of these roads by unloading the road trucks there. This was often far from popular with the drivers of the trucks, and I shudder now when I think of some of the places we induced them to reach with a load. I am glad to say we never had a serious accident.

From that point on it was pack-horse and hand spreading. The first hill block I topdressed was one of 150 acres, its nearest point about a mile from the homestead. Here we made a dump of fifteen tons of North African Phospate, shipped from the port of Sfax in Tunisia in the Mediterranean. I had engaged two men to do the spreading by hand, while I proposed to keep them supplied with a team of four pack-horses. When they realised I meant to do this single-handed, they nearly jibbed, as they thought I would be unable to keep them going. In the event I packed the stuff out to them faster than they could put it on. Beside the dump I had dug a trench into which the horses could be led one by one. A cornsack was hooked on to the packsaddle hooks by its two bottom corners. The horse was led into the trench, the top of which was level with the bottom of the pack-saddle. This allowed half the sack on each side to rest on the ground. On to this one each side I rolled a 140-pound bag of phosphate. This was quickly rolled tight against the saddle by picking up the outer corners of the sack and hooking them on to the saddle hooks. One packstrap was tightened round each bag to stop it flopping about.

I found I could load my four horses, all good sturdy types, with eight bags, half a ton in all, in a matter of minutes. Riding another horse, with a whip and a spare rein, I then drove my team out to the area where the spreaders were, where I caught each horse in turn, snapped on the spare rein, and led him to the appropriate spot to dump his load of two bags. The empty carrying sacks then securely strapped to the saddles, I got behind my team and drove them back at a good clip to the loading point. On some of the easier country we were able to sledge the fertiliser out from the dump half a ton at a time. This was for many years the only means we had of getting fertiliser on, and we persevered with it, trying to spread some 20–30 tons annually. I took my turn in staggering round the hill-sides with a bag holding the best part of a hundredweight of manure draped over my protesting frame. It was exhausting work.

Then some bright lad came up with a New Zealand invention, the fertiliser 'blower'. This consisted of a hopper holding about 150 pounds of fertiliser, from the bottom of which it was fed down in front of a high speed engine-driven fan, which blew it out under considerable air pressure through a long spout which, the whole outfit being on a turntable, could be turned in any direction. This worked quite effectively with the material we were then using, North African phosphate, which was very finely ground. Making use of light air currents, the stuff could be blown in a dense cloud for many chains.

This appliance was a godsend, as it took the desperately hard work out of this job, for which it was increasingly difficult to get men. The obvious way to exploit this new method was to put in more bulldozer roads. This we did over the years, till we ultimately had about eight miles of roads on the farm. I bought a war-surplus Dodge four-wheel drive truck, to which I had a winch with 250 feet of wire rope fitted. We bolted the blower down on the truck, which could carry a ton or more of fertiliser with it. We had dumps made by the road trucks as before, and operated from these over the newly made bulldozed roads, these often leading on to the tops of spurs. When we got stuck, we hitched the cable on to a substantial stump or puriri tree, and pulled ourselves out with the winch.

As I write I can see some of the high points we reached with the truck. One stretch was known as the 'Burma Road'. It would give me 'the willies' to have to take a wheeled vehicle to some of those places today. The blower did not give a full coverage, owing to the difficulties of terrain, and the spreading was uneven, but it was a big advance on hand spreading, and produced the first decisive upturn in fertility of our hill pastures. My son was now helping me on the farm after obtaining his Diploma in Agriculture at Massey College. He played a big part in all this work of improving the land, particularly in 'logging up'—that is, pulling together in heaps for burning, half rotten and useless timber cluttering up the ground. This was done with two good draught horses and a helper.

As visitors survey New Zealand's topdressed green hills, millions of acres of them, that have been developed from bush land, how few can have any inkling of all the effort that has gone to the making of this vividly green sward! I tend to react sharply when someone says 'How "lucky" you are to have a place like this!'

Chapter 20

THE WAR YEARS

Those of us who had taken part in World War I hoped that there would never be another similar conflict, but as the portents of such a possibility became more ominous in the thirties, many wondered what course they should take personally towards the avoidance of war. All kinds of peace and anti-war movements attracted their share of opinion, all alas to be in vain in the inexorable march of events. For a long time I wondered if perhaps the pacifists had the right answer, until I came upon the phrase, somewhere in my reading, 'a condition of justice is resistance to injustice'. This seemed to me to be a complete answer and refutation of the pacifist case. One was then left with the only answer, which was that the tyrant must be resisted, by physical force, and all it involved, if necessary.

The years of World War II—1939 to 1945—brought strenuous and sad times to New Zealand, though we must count ourselves fortunate that our country was not invaded. Initially we undertook a responsibility in supplying forces for the Middle East and their reinforcement. Then when the Japanese entered the war we had to find troops for the Pacific theatre and for the defence of New Zealand itself. As a comparatively small community of, at that time, under 2½ million people, this strained our resources in man-power to the utmost. With all this we had to see that our essential industries, in which farming ranked high, were not crippled. We older men, many of us veterans of the first world war, had to do our best by extra effort to replace those called up. Farm labour, never plentiful, became almost impossible to get, and our splendid farm wives helped us out with many of the jobs that were difficult to do single-handed.

A neighbour went off to the Air Force, and I supervised his farm as well as running my own; Many others did the same. Rationing of essential items, particularly petrol, did not make things easier, but everyone 'hoed in' as we say, and made the best of it. I was at that time a Dominion Vice President of the New Zealand Farmers'

Union, which involved others and myself representing the farmers in discussions with Ministers on all sorts of matters concerning the efficient running of our industry under war-time shortages. When the Japanese entered the war, New Zealand, with most of her mobilised manpower in the Middle East, was almost defenceless, had any long distance seaborne attack been made on us.

I was one of a deputation of four representing the New Zealand Farmers' Union, all of us ex-soldiers, which waited upon the government to urge the recruiting and arming of a Home Guard to supplement the forces being mobilised for New Zealand defence. We were received by the Prime Minister, Mr. Fraser, and other members of the cabinet. Also present was the Chief of Staff, General Puttick, who on a strictly confidential basis, with the aid of a large map, briefed us on the actual position in the Pacific area at that time. We were also told how short we were of weapons. We subsequently learnt that a former government had sold the complete equipment for a division to the South African Government during the '30's. The picture painted was horrifying, showing that we were far more defenceless than the public imagined. I found it hard to keep all this secret stuff under my hat on the long journey home, after the meeting, by train and bus to the northern peninsula where I lived. We had been told that this area was particularly vulnerable and might be cut off in a seaborne attack.

The government lost no time in authorising the setting up of a Home Guard and this was one more activity that had to be regarded as urgent. Practically every able-bodied man in the country areas volunteered, and we had a hectic time getting organised into units and trying to scare up weapons. I commanded a company drawn from a rural district about twenty-five miles long by six or eight wide. This produced two oversized platoons and one mounted rifle troop. We started by making a supply of 'Molotov cocktails' and dug up every sporting rifle and shotgun in the district. We had a mould made for casting solid lead bullets for the 12-bore shotguns. I am glad to say that old army Lee Enfield magazine rifles became available before we had to do any range practice with these.

While waiting for rifles we did what drill and instruction we could, and dug defensive positions commanding road junctions and bridges. When rifles arrived our morale soared, and we began to feel we could play a worthwhile part if the balloon went up. My concern, as an ex-musketry instructor in World War I, was to qualify as many men as marksmen as I could. Apart from the few men used to sporting weapons, and ex-service men, it was surprising how many had not fired a service ·303. After lots of instruction on the rifle, particularly with aiming practice, came the day when live ammunition arrived, and we could start actual firing practice on the range.

We located areas suitable for 25-yard ranges using small targets where there seemed little danger to others. I was not so sure about the safety factor after the first range practice, particularly with the large Maori platoon of over forty men. The targets were at the base of a terrace I had thought high enough to meet all eventualities. When my warriors commenced firing I had a few shocks. The sighting of the oldish rifles was often defective, and many of the troops were far from confident. Consequently bullets were kicking up the dust from five yards in front of the marksman to half-way up the terrace, and the whine and 'whee' of ricochets was heard far too often for my liking. At lunch-time I sent a man off to move his family out from the nearest house beyond the range as a precaution.

We ultimately got this sorted out, with the rifles sighted properly, and my Maori platoon became a formidable unit that would have scared the day-lights out of any Japanese had they encountered them. They loved any ceremonial drill such as guard duty, and presented arms with a snap and precision that would have done credit to any regular unit. Gradually more gear became available, and the Home Guard, with equipment and training, became a force that would have given a good account of itself had the necessity arisen. My men knew every track and fold of ground in our vicinity, and in tactical exercises had the natural aptitude of countrymen for making the best use of ground. There was plenty of improvisation and a good deal of humour. Ill though we could spare the time, we made the most of the periods given to training.

In addition to the Home Guard we also made our provision for the possibility of the northern peninsula being cut off by an enemy landing north of Auckland city which would have resulted in enemy occupation. We arranged a camp site and organised the necessities for it such as blankets, cooking gear, packsaddles, reserves of food and so on. This was in the middle of the bush concealed from air observation, and so remote that it would be hard to find. The intention was to concentrate our women and children in this spot, where with our superior bushcraft, we thought we could scupper any forces attempting to find them.

Our doings on the home front were of minor importance compared to the world shaking events and tragedies that were occurring overseas, but we did at least prepare to defend ourselves, and kept up production of the supplies so urgently needed by Britain. The latter was no mean feat.

There is a tendency by some to make light of all this effort, but let it be remembered that it was deadly serious at the time, when there was literally nothing but distance between us and the, at that time, victorious Japanese. The arrival of American forces was a great occasion, and made us feel we were no longer alone; but it was not until the sea battles, first of the Coral Sea, and then Midway, that the possibility of invasion really receded. The masterly tactics of Admiral Nimitz in the battle of Midway, in which he crippled the Japanese carrier force, was a decisive event for New Zealand. We in this country should never forget what we owe to him, and the other great Commander, General MacArthur, in driving back and defeating the enemy forces that threatened Australia and New Zealand.

When New Caledonia became a base for forces in the South Pacific, aircraft were shuttling back and forth between Auckland and Noumea, their route being directly across our locality. One day two of us were drafting a mob of Hereford cattle in a small hilltop paddock when a 4-engined super-fortress came roaring over the sky-line and straight towards us. As he circled round I felt sure the pilot must be a home-sick cattle man from Texas or Wyoming, and could almost hear him saying to his crew 'Well what do you know—Hereford cattle just like home—let's have a look at what these guys are doing'. And have a good look they did before setting off on a course for Auckland.

An area of 5,000 acres of gum land not far away was used for artillery practice, and the Americans provided us with a minor sensation. Firing heavy guns, they were using part charges for practice, but some mistake resulted in the firing of a full charge, causing the shell to travel well beyond the target area and on to a local farm. The heavy shell burst almost under the main powerline to the North, cutting it, and killing a dairy cow, and frightening the wits out of a neighbour's wife picnicking nearby with her children.

Chapter 21

AERIAL TOPDRESSING—NEW LIFE FOR THE HILLS

Although it is running ahead of my story chronologically, I will complete the picture of the topdressing theme and bring it up to date. For many years I used to look up at the steep hills at the back of my farm running up to 1,500 feet in altitude, and wonder what I could do about them. Erosion was showing itself, and the fertility of this land was declining, but it was not possible with any means then at hand to give it the phosphate it needed.

The answer to this became apparent when it was found that aircraft could play a prominent part in farm mechanisation. It was the development in aircraft and the greatly increased interest in flying due to the war that really set this off. Long before the war some of us had seen the spreading of fertiliser from the air as the ultimate answer when suitable aircraft became available. Pressure for experimental work to be done was put up to the Government by farming bodies, and we were fortunate that at that time there were some senior officers in the R.N.Z.A.F. who were imaginative and interested in the idea. A dramatic series of trials took place in the Wairarapa in April 1949. During the three weeks of these experiments, in the words of the officer in charge, 'We dropped 137 tons of Super and Lime, and hit the targets on most occasions'. The aircraft used were three Air Force 'Avengers' plus an Auster.

As this was going on, private firms were not slow to see the possibilities in this new development, ex-war pilots were interested, and war surplus Tiger Moth aircraft, which had been widely used as training aircraft in the war, could be bought at reasonable prices, about £400 a time. Experience and enquiry showed that the job called for special qualities in an aircraft rather different from those possessed by the R.N.Z.A.F. planes that had been used in the experimental work. The requirements for an aircraft to be used in topdressing hill country were assessed as below:

(1) It must be capable of taking off fully loaded, from a paddock or grass strip 3–400 yards long.

(2) For accurate spreading it must be able to fly low (40–80 feet) and slow (50–100 m.p.h.).

(3) It must have sufficient manoeuvrability to be able to make sharp turns, fly around hillsides and in and out of steep valleys.

It is interesting to note that many years after these requirements were first accepted, and when we are using much better aircraft, they still hold good.

At that time the Tiger Moth or DH82 was the only aircraft available with the necessary qualifications. It was cheap, and available in large numbers. Built by De Havillands in 1926, and equipped with a Gipsy Major engine, it was the Model T Ford among aeroplanes. It had two seats, one in front of the other, and fitted with dual controls it could be flown from either seat. It was an ideal training plane and for twenty years most of the pilots in Britain and the Dominions were taught to fly in 'Tigers'. After the war they were available in hundreds from 'War Assets' and it is amazing that, after completing twenty-three years of useful life, these 'kites' of sticks and fabric became the backbone of a new industry, while all the modern war aircraft were left to rot or melted down for scrap.

It has been estimated that the Tiger Moths carried over a million tons of superphosphate, spreading it over seven million acres of farm land, nearly all rough hill country, during the years they were used for topdressing in New Zealand. At this point I think it is appropriate to take a bit of credit for New Zealand initiative in creating a new technique in which we led, and still lead, the world. It was a joint effort between the farmers, our splendid New Zealand pilots, and the private firms which were formed to do the work, and which provided a great example of co-operation, with excellent relations between all those concerned.

It was the sheep farmers, farming our hill country, who were mainly interested. Seeing the immense possibilities in this new idea, they were ready to do their part, which consisted in providing suitable airstrips, with road access to bins which would hold bulk fertiliser at the loading point. A typical cost would be say £350 for earth formation work for the strip, and up to the same amount for a bin, according to capacity. These airstrips could not be put just anywhere, they had to be properly sited from the flying point of view, upon which most farmers were glad to be advised by pilots.

These men, the pilots, mostly ex-air force men, did a great job, and quickly evolved techniques which enabled them to make a reasonably even spread of fertiliser from the air over all sorts of rough country. It was quickly apparent that here lay the answer to the rejuvenation of our steep land over millions of acres in the North Island of New Zealand. The independent private firms who undertook this work deserve a lot of credit. Having bought one or more Moths, they had to undertake the costly job of converting it from a training aircraft to a topdressing aircraft. This involved stripping the front cockpit, with removal of all gear, fitting it with a manure hopper, and then connecting everything up again in such a way that structural strength and balance were unimpaired. The total cost of all this ran out at about three times the original cost of the aircraft!

This, however, was only half the job. Experience had shown that if this work was to be done at a reasonable cost, quick loading and turnaround of the aircraft was essential. With typical New Zealand inventiveness, a hydraulically operated hopper was designed which, at the end of long arms, could be worked from a tractor or truck. Laid on its front it acted to scoop up the fertiliser, pulled upright and lifted high by the arms it was ready to be driven up behind the aircraft and release its load through a canvas tube into it. These loaders were steadily improved, until the standard unit today is a combination of fuel tanker and loader costing several thousand pounds.

A pleasant feature of all this development of a new industry was the good relationship between the topdressing firms, the pilots and the farmers. As one of the latter, I can say that we truly appreciated what the work of these men, and the risks they took, meant to our industry and country, and we and our wives were glad to do anything we could to help. I think the pilots and loaders liked the meals our wives put on for them. The firms found the farmers good customers, and one of the largest reported that after ten years of operation they had not incurred one bad debt.

The Tiger Moth performed amazingly, but as the industry grew and replacements and maintenance of already aged aircraft became heavier, it was obvious that some type of aircraft specialised for the work would have to be sought. World wide enquiry was made for this, the need being ultimately met by a machine built by the Fletcher Aircraft Corporation of California. Although other makes continue to be used, the Fletcher is now the work-horse of the industry and has proved itself to be well up to requirements.

A limitation of the Tiger Moth was that this light aircraft could not take more than 500 pounds as a load, even this being a remarkable evolution for so light a machine. The Fletchers can take off comfortably with a load of fifteen hundredweights, three-quarters of a ton, which means a big increase in the tonnage spread per flying hour, the significant factor in the costing of the job.

As financial considerations allow, farmers are tending to step up their application of phosphate to the hill country, which is carrying more and more stock, our sheep numbers now being well over the fifty million mark. As I write, the last recorded figure for tonnage applied from the air was over 800,000 tons. It is expected to be around 1,000,000 tons this year. Through my office window I can see the pilot taking off from my airstrip a few chains away almost monotonously, carrying three-quarters of a ton each time as he lifts into the air with his engine howling at full boost. He has 170 tons to spread on the hills.

I reflect what an advance this is on the days when we used to stagger round the hillsides carrying three-quarters of a hundredweight, and working ourselves to a standstill to apply perhaps twenty-five tons. All hail to the topdressing pilot!—he is a splendid fellow, and deserves some kind of accolade for the vitally important job he does for the economic welfare of New Zealand.

He deserves something else too, and that is more assistance financially in qualifying himself for the job, learning to fly is an expensive business, beyond the resources of many young men who would otherwise take it on. Although the State does provide some assistance in this, it is not nearly enough, having regard to the importance of this profession to our exports, and that many of these pilots ultimately become pilots of passenger aircraft. Also the best protection against accidents is thorough training, especially with loaded aircraft. More liberal financial help would ensure this and save lives. What about it, Mr. Minister for Civil Aviation?

It was in the early 1950's that aerial topdressing came into its own. These were the golden years for my generation of farmers. Prices for wool and stock were good, and surplus income was readily spent in forming airstrips and from them then applying from the air the large quantities of phosphates that our hill country so badly needed. There was a general upsurge of production from which the whole country benefited.

Chapter 22

SELLING OUR WOOL

Our industrial age has sought to lessen man's dependence on natural products derived from animals, plants and minerals. In this the great chemical corporations have played a leading part. They finance extensive research on the synthesis of new materials, and have had considerable success in displacing traditional fibres and foods in particular. Examples are artificial silk, margarine and a whole range of synthetic fibres.

The synthetic fibres constitute a challenge to wool and have invaded many of its markets, but have not succeeded in pushing wool into the background. Lavishly advertised as 'miracle fibres' they have often failed to live up to the promises made for them, and none produced to date has the many qualities of the natural fibre of wool, designed by nature to protect and clothe the humble sheep in a vast range of environmental conditions. The astonishing thing about wool is its versatility. It can be spun, woven, knitted, dyed and felted. It produces an immense variety of clothes from rugged tweeds to the most delicate fabrics worn in the ballroom. It supplies not only apparel, but curtains, carpets, upholstery and a wide range of felts. It does all these things well because of its inherent qualities. It is warm and comfortable, with good properties as an insulator against heat and cold.

Wool is absorbent and removes perspiration moisture from the skin. It will absorb up to 30 per cent of its weight of moisture without becoming damp or clammy, in strong contrast to the cold damp feel of a synthetic garment worn next to the skin. This valuable property makes for good hygiene, and affords a protection against chills. It is strong and durable and has life and elasticity. Its resistance to fire makes it safe wear for children in particular it does not flare up or melt as do some of the synthetics. Wool takes dyes readily, and this combined with its good draping qualities makes wool fibres the preferred ones in the top ranges of colourful fashion garments.

I hope I may be excused this panegyric on the fibre which I have been engaged in growing for the last fifty years. The reason for it, and its relevance, will I hope become apparent in what follows.

As far back as 1937 it was becoming apparent to woolgrowers that man-made fibres were soon going to challenge wool in markets that had been regarded as its own. In that year representatives of the sheep farmers of Australia, South Africa and New Zealand met in Sydney to consider this threat. The outcome was the setting up of the International Wool Secretariat in London, for the purpose of fostering and promoting use of wool in the consumer countries. A secondary function was to foster research on methods of improving the performance of wool, such as processes for resisting shrinkage. Funds were provided pro rata by the three wool-growing countries by a levy on each bale.

Commencing its work on a modest budget of a mere £50,000 a year, the activities of the I.W.S. as it has come to be called, were soon interrupted by World War II. As the war progressed and the Germans over-ran Europe, a fresh alarming problem in wool developed. On the outbreak of war, Britain had commandeered the wool clips of the three British Dominions, the intention being to supply her Allies as well as herself with the wool needed to clothe millions of troops. As her Allies disappeared one by one into the maw of the German military onslaught, Britain was left with more wool on her hands than she knew what to do with. The continental peoples on the other hand were desperately short of wool, and had to shiver in cotton and various types of synthetics through their bitter winters. As the end of the war drew near there were no less than ten and a half million bales of wool piled up in storage. It was obvious that if this quantity was released without any regulation that it would wreck the wool growers' market for their oncoming clips, and push many of them into bankruptcy, with detrimental effects on the economics of the wool producing countries. The growth of wool cannot be shut down as can a chemical fibre plant.

The U.K. Government invited Australia, New Zealand and South Africa to co-operate with it in working out some means of realising all this wool over a period so as not to wreck the market. Each Dominion enlarged the body that had hitherto coped with wool promotion, and gave it the extra powers needed to deal with the new problem. These were known as Wool Boards, and had a predominance of elected wool-grower representatives in their make-up. In New Zealand the setting up of the Wool Board took place in 1944, at which point I personally became involved in all this activity concerning wool.

As a national Vice President of the farmers representative body, at that time the New Zealand Farmers' Union, I had been active in trying to get something done about what seemed to be a pretty desperate future facing our sheep farmers. I was elected as one of the six woolgrower members on the Wool Board, and we were immediately told to get on with the job of planning to liquidate the immense stockpile of raw wool. We chose delegates to meet with others from Australia, South Africa and the U.K. in London.

After months of discussion there emerged what came to be known as the 'Joint Organisation' or 'J.O.' which was charged with feeding this wool to the markets of the world over a long period, estimated then to be thirteen years. None of us realised what the demand for wool would be. Completely starved for five years of the imports of wool on which they had customarily relied, and having had to make do with inferior substitutes, the cry from Europe was 'for God's sake let us have genuine wool again.' Large quantities found their way to continental manufacturers, and once again millions of people had the comfort and warmth of wool which had been denied to them for so long. In the event, such was the demand for the natural fibre that the stockpile of ten and a half million bales was put into consumption in seven years instead of the estimated thirteen, and without seriously upsetting the market. It was a triumph of intelligent anticipation and good organisation—without it there would have been chaos.

During the immediate post-war period as billions of dollars were poured into the ex-enemy countries for reconstruction, the synthetic threat to wool loomed ever more ominous, as new factories were built and large sums were spent on research.

This was the cue for the woolgrowers to intensify their efforts, by imposing heavier financial levies on their production to pay for a rapid extension of their efforts in promotion and means of improving the performance of their fibre. Thus began, for me, a period of twenty years service as a representative of the woolgrowers on the New Zealand Wool Board, during which time I retained my seat at successive elections until I retired in 1964. This was an extra-curricular job in addition to that of being a sheep farmer developing a large property, which often strained my mental and physical resources. It was an illuminating experience involving me in many interesting and colourful events, and the odd frightening one, a few of which I relate.

Living, as I did, not far from the northern tip of New Zealand, travelling once or more a month to Wellington, the capital, where most of our meetings were held, was quite a thing, a journey of 500 miles each way, and flying was the only practicable way to do it. The service between my local centre, Whangarei, and Auckland, about 100 miles, was provided by De Havilland Dominie aircraft, a two-engined biplane seating six. At Auckland I joined the main trunk flight to Wellington provided first by DC3's and in later years by Viscounts, one of Britain's outstanding successes in aircraft. In winter months the weather was often foul, and I had one or two 'dicy' experiences of 'all weather flying'. On one occasion a north-easterly gale intensified after leaving Auckland, where we had been strapped into our seats in the hangar before the aircraft taxied out into the wind. Most of the route lay over the ocean several miles off the coast, where we felt the full force of the rising gale. As violent gusts of wind and heavy rain struck us like a heavy hand trying to push us back, I thought that at any moment we would be dunked into the whitecaps of the raging sea a few hundred feet below us. When we landed never was I more glad to set foot on terra firma again.

Once our local grass airfield was covered with sheets of water from a recent heavy downpour. The pilot duly taxied us out through this to the take-off point. There he turned back to his somewhat apprehensive passengers and announced reassuringly—'Don't be alarmed if I don't get unstuck on the first go.' Thereupon he gunned the engines, and with spray flying up all round us, just managed to stagger into the air at the far side of the aerodrome. On yet another similar occasion the aircraft became bogged down and a tractor had to be requisitioned to pull it out. Once I flew to Australia in one of the flying boats that in those days provided the service. It was a long flight through stormy weather, and took nine hours as against the two and a half hours of the modern jet. High over the Tasman sea in the darkness there was a blinding blue flash on the port wing, and a crash like a heavy shell exploding. I thought the wing must have been blown off—we had been struck by lightning, but apparently without damage, as the four motors continued their reassuring song. One had to learn to take this sort of thing in one's stride as part of air travel and become something of a fatalist.

The work of the Wool Board was divided between activities of all sorts in New Zealand, and the larger sphere of international

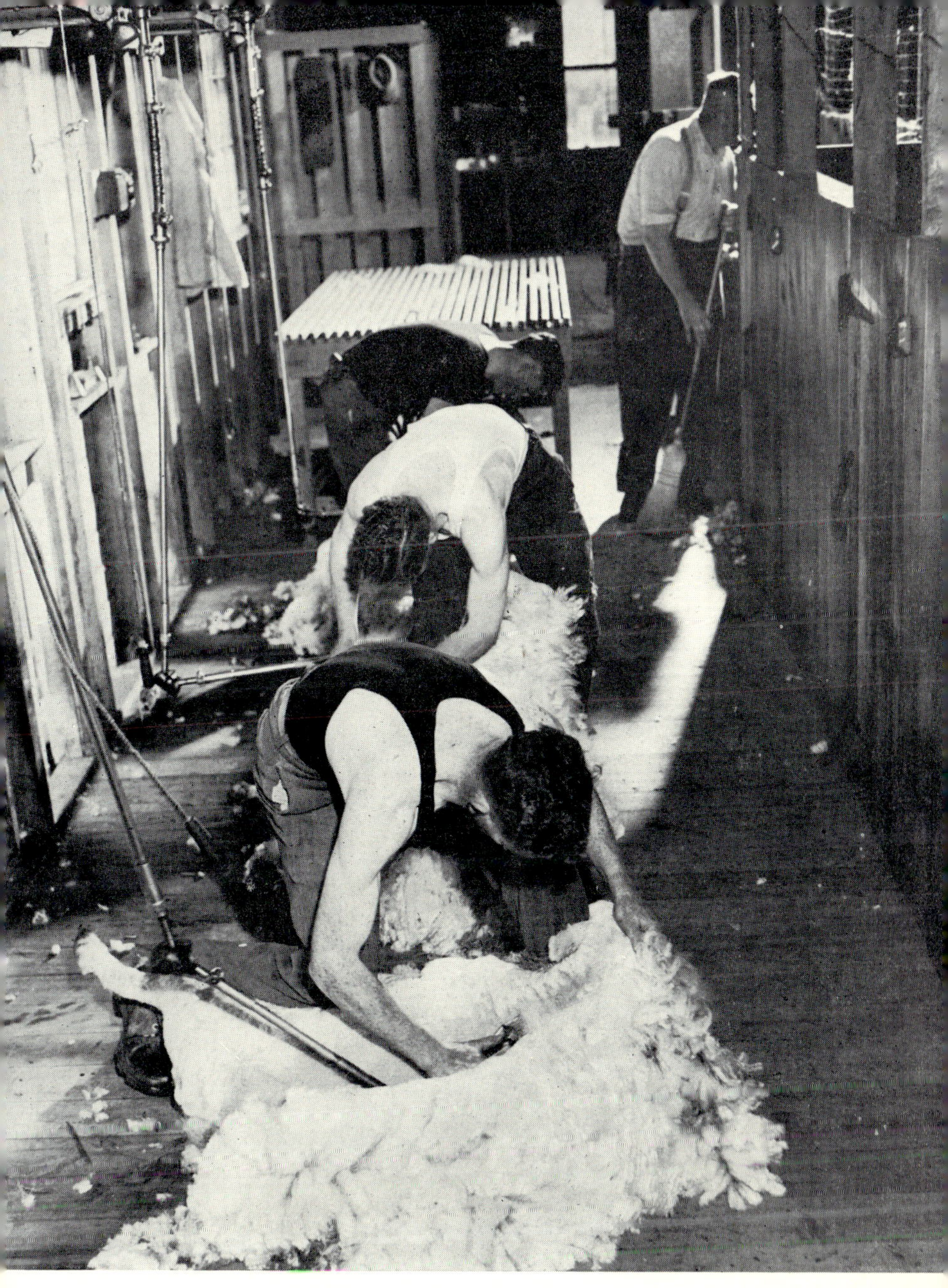

Harvesting the year's clip of wool. Shearers at work.

Lambs being held for docking—time to get rid of those tails!

Fine-woolled sheep.

wool marketing overseas where most of our funds were spent. Fashion shows, indicating what could be done with wool at its best, have always been one means of promoting wool, and were popular with women. We men had to learn something about this, since from time to time we had to do our stuff in opening such functions. I found it took quite a bit of nerve, and some rather false assurance, to stand up before hundreds of women with hardly a man in sight, to tell them what it was all about. Once, searching for the back entrance to the stage to make the opening speech, I inadvertently barged into the models' dressing-room. They seemed to think it was all part of the fun! Incidentally the assurance and poise of these girls who showed off so gracefully some of the beautiful fashion creations always amazed me. Nowhere was this better displayed than when we put on a super show of wool fashions before the Queen on her last visit to New Zealand.

Because we spent most of the growers' funds overseas, we had to ensure that they were spent wisely and to good effect in encouraging the use of wool. This necessitated extensive travel by members of the Board, usually two at a time. We met our Australian and South African contemporaries in the London conferences which settled policy, and with our American executives of the Wool Bureau in New York. En route we were able to see something of the activities of the directors and their officers in various European countries.

Chapter 23

INTERNATIONAL TRAVELLER FOR WOOL

The journey that stays most vividly in my mind was one during the autumn of 1956. The Board Chairman and I flew to England via America, stopping off in New York for conferences with the American executives of the Wool Bureau there, the body responsible for the expenditure of growers' funds to popularise wool in U.S.A. After a fortnight of concentrated discussion in London, we visited the continental offices of the I.W.S. in Paris, Milan and Dusseldorf. Just out of this German city we were taken to see a very modern mill. As we sat in the manager's office, being told how the mill had been rebuilt after the former building had been bombed flat, my attention was drawn to a portrait of a fine-looking man on one wall. Dramatically the speaker stopped and then said: 'Our former Chairman of Directors, hung by Hitler'—and continued his discourse without any further allusion to the subject. From Dusseldorf we flew to Copenhagen.

The Scandinavian Air Services had recently opened the trans-polar route to Tokyo, and we had been instructed to return to New Zealand via Japan, with which country we were doing a steadily increasing trade in wool. Copenhagen was an entrancing place, with its quaint buildings, many of them embellished by towers sheathed in copper, and its friendly Danish people, some of whom spoke English. We left Copenhagen at 9 p.m. at night in a DC7B, and flew for two hours or so to Lulea, at the head of the Gulf of Bothnia, the port to which in the summer time is railed the iron ore from the mines at Narvik on the north west coast of Norway. After a brief stop, just long enough to top up the fuel tanks of the aircraft, we took off on a route directly across the North Pole for Anchorage in Alaska. We had bunks, and I slept fairly well.

The following day was a brilliantly sunny one, with hardly a cloud in the sky, and the sight from the aircraft was an amazing one. As far as the eye could see on every side from horizon to horizon, was glittering ice, piled high in broken ridges and hummocks. Leads of water showed every here and there in the cracks between the piled masses and floes. It was an eerie desolate place. At 10.30 a.m. by Copenhagen time, by our watches, came an announcement from the captain, made in three languages. It was: 'We are now at an altitude of 13,000 feet, and directly over the area of the North Pole.'

The three languages were necessary, for this was a truly international journey so far as people were concerned. The crew comprised Swedes, Danes, and Norwegians—the hostess, I remember, was a charming girl from Oslo. Amongst the passengers were Germans, French, Dutch, English, Chinese, Japanese and Scandinavians. A very sedate elderly Chinese who was obviously a person of some consequence sat across the aisle from me. I have no idea who he was, but I was amused, later, in Tokyo, to see the customs officials going through his effects with a fine-tooth comb, suggesting that perhaps he might have been less respectable than he seemed.

Alaska proclaimed itself as we droned on with a distant view of Mount McKinley looming above the clouds, the highest mountain in North America. After what seemed a long day, about five in the afternoon by Copenhagen time, we landed at the airport of Anchorage, to be told that it was a quarter to six in the morning by their time. After that I gave up trying to understand the vagaries of the time scale.

No snow and ice around close at hand as one had somehow imagined, but a modern airport with dozens of small aircraft, extensively used in a land rather short of roading. It seemed hard to believe that a city of 90,000 people existed a few miles away in this remote spot. One hour on the ground sufficed to refuel and service the aircraft, during which we were closely shepherded into a transit passenger's lounge by a huge armed cop, who was no doubt responsible for seeing that no doubtful characters escaped into Alaska. During the checking of passports the immigration official commented on our two from New Zealand, saying they were the first of their kind seen on that route. Perhaps we were the first two New Zealanders to cross the North Pole.

Off again into the air for Tokyo, the beginning of a new day somehow telescoped into the old one without the intervention of a night and the sleep that usually goes with it. The day did seem interminable, and we had just about 'had it' as we came in to land at Tokyo about four on a murky afternoon, after a total of thirty hours in the air. Our International Wool Secretariat in Tokyo had worked out a programme for the eight days we were to be in Japan. This enabled us to meet many of their top wool men, and to see a number of their mills and something of the land and its people. It was all intensely interesting, and we were impressed by their considerable knowledge of everything to do with wool and their very well laid out mills, many of them rebuilt after the war with American aid.

Japan is far from being a tropical country, and can be very cold, about half the country being visited with snow every winter. The heating arrangements in their lightly built houses are not very effective, and they are appreciative of the warmth of wool. Their standard of living is rising fast, due to hard work and good planning, and western dress is becoming almost generally worn. Accordingly they can afford good apparel, and Australia and New Zealand are very glad to provide them with all the raw wool they need. Their own sheep flocks are comparatively insignificant, about 207,000 (1965) in a nation of over ninety million people, compared to New Zealand's fifty-nine million in a nation of two and a half million people.

There is no question of sweated labour in the textile industries of Japan today. The industrial unions are strong, and able to demand good wages and conditions for the workers in their mills. In one large plant we were courteously received. The office staff bow as you pass them to a small reception room. There the visitors are regaled with refreshing green tea, made from the tips of the young leaves, and each guest is provided with a tightly rolled small towel in a saucer. This has been wrung out from hot water and is used to wipe face, neck, and hands of perspiration. We found such to be the invariable procedure in visiting a mill. At this session the management produced statistics of the enterprise, and there is an interchange of questions. They had few complaints about New Zealand wool, many more about Australian wool, chiefly to do with 'second cuts' where the beautiful merino staple is cut and chopped about by bad shearing, and skin pieces which had to be picked out.

They told us that 80 per cent of their employees were girls, with keen competition for the job when a girl dropped out to marry. The girls lived in dormitories, had an excellent cafeteria and television theatre, besides various educational facilities, and were paid wages that seemed good having regard to their price level. We were in the mill buildings as the staff broke off for lunch, and a mass of laughing girls like a flock of birds swept past us; there could be little doubt they were happy in their jobs. The Japanese low cost of production is due to sheer efficiency. Their stocks of raw material are well bought, their factories designed for the best lay-out, and equipped with good machinery, most of it made in Japan. At this mill they had two complete staffs, each working a forty-eight hour week. Their machinery was thus kept running sixteen hours a day or ninety-six hours a week. Our forty hours a week less tea breaks could never compete with this for efficient production.

We saw a variety of mills and processes. The kimono is still used for dress wear by Japanese women, and although it cannot be a very comfortable garb, they manage to look very charming in this traditional dress. One interesting factory we visited in Kyoto specialised in the design and making of high-grade kimonos for the well-to-do. They had their own design artists, four of them, one poor chap without arms painting with the brush in his mouth. Only a few garments of each design were made, to retain its exclusiveness, and they retailed for £25–30 each. They had three Jacquard looms, on which they reproduced in natural silk brocaded on the finest of merino wool fabric, the most exquisite designs and colours. I found it hard to tear myself away from all this.

The Japanese are an artistic people, honest, courteous, and hard-working, but with a certain inscrutability the westerner finds hard to fathom. These qualities are hard to reconcile with the dreadful things their army did in the war, and it was not until by dint of careful questioning I found that a large proportion of the population hated the militarists and all they stood for, that I got some understanding of the apparent paradox.

Much of the wool textile industry is centred at Nagoya, 80 per cent of which was reduced to ashes and rubble by wartime bombing. In the course of rebuilding the city a far-seeing civic authority has widened the main streets and planted them extensively with ornamental trees, so that the new city as it grows should become a place of beauty. The Japanese do not normally entertain visitors in their own homes, and one seldom meets the wives of the business men you deal with. They are none the less hospitable, and a usual way of entertaining overseas guests is to take them to a geisha tea-house for the evening meal and entertainment. These places specialise in this kind of thing, charges being quite high. We were entertained by a top figure in the wool business, and gathered in the early evening at a tea-house in a pleasantly rural setting in a miniature garden with a stream, which the Japanese contrive so well. After removing your shoes on entering the building, you are accommodated in a rather uncomfortable squatting position around a low-set table on which the meal will be served.

A geisha, resplendent in traditional dress of beautiful kimono and obi, with hair ornaments, is seated beside each male guest. She may not be particularly good looking, but everything she does is done with grace and charm. After an apprenticeship of three to four years she is a highly skilled entertainer, and it is her business to see that the guests enjoy their evening. She sees to it that your small cup of hot saki, or rice wine, is kept filled, and encourages you to 'drink up' with this or the beer that is increasingly displacing it. A succession of courses of small tit-bits is placed before you, every item exquisitely served and garnished so that it seems a pity to eat it. They feature of course traditional fare which a westerner finds hard to identify, though I was pretty certain that pieces of raw fish I tackled were just that, in spite of its disguise. I was to learn later that one of the other unidentifiable courses was raw octopus!

After a leisurely meal with a good deal of laughter and backchat, helped by the effect of repeated doses of saki, the 'mama san', or manageress, produces a samisen, a banjo-like instrument with strings. The geishas take it in turn to play and sing to you in graceful little action songs in which as much art goes into gestures as the rather plaintive little melodies. If the party has warmed up properly the girls then are ready to dance, and you are invited to join in. This may be followed by more or less polite parlour games, depending on how far the usual inhibitions of the guests have become neutralised. At the appropriate time, clear indications appear that the evening's entertainment is at an end, and you are charmingly bowed out to put on your shoes and return to your hotel.

At Osaka, known as the 'Venice of Japan', on account of its many canals and 1,200 bridges, we were entertained to lunch by representatives of sixteen importing firms that buy and bring into Japan 90 per cent of all the wool used. These were alert men of the world, familiar with the buying centres of Australia and New

Zealand, most of whom spoke English. Here we had one or two complaints about specific lots of New Zealand wool which we were able to take up with those concerned on our return home. After eight days travelling in Japan we were able to get a good view of their wool industry and see what our wool was used for. One of the astonishing things was to find that 25 per cent of all imports went into hand knitting wool—not so much for knitting by hand as for use on home knitting machines. These were available, Japanese-made, at a fraction of the price ruling in New Zealand, and we were told they were widely bought by women's organisations who shared them round. As one man said 'they run up a pullover in an evening'. We were very interested in this, since fine Southdown cross wool of 58's quality was in strong demand for this purpose, a type we can supply.

The Japanese never lost an opportunity to emphasise the necessity for reciprocal trade. They knew all about the controversial proposal for establishing a cotton mill in New Zealand, and said to us in effect—'why manufacture cotton goods in your country when we can sell you all the finished goods you need at about half the price?'—a truly incontrovertible argument with which we, as farmers, were in agreement. New Zealand's trade with Japan is growing steadily, but can reach much larger proportions. They can take from us primary produce of all sorts in volume, in exchange for an extensive range of manufactured goods of good quality. If we are foolish enough to allow certain interests to throttle this trade by excessive protective measures in favour of our small scale manufacturing industries, we will suffer a drop in our own standards of living.

After our stay in Japan we returned to New Zealand by air via Honolulu, having successfully survived the Asian 'flu which was rampant in Japan at that time. In our two months away we had travelled 30,000 miles. The reader may well ask what place this sort of thing has in a farmer's life, and whether it is really necessary. The answer is that farmers have found by experience that it is not enough to confine their activities to production, and throw their meat, and wool, and butter on the world market for others to deal in. They have for many years past elected their own representatives to attend to such things as orderly marketing, maintenance of quality standards and improvement of the product, and sales promotion in the big consumer markets. They do not hesitate to levy themselves to pay for this, which has now become a permanent part of our New Zealand farm system, and has produced significant results, particularly in the widening of our markets. A recent table of ultimate destinations of our wool listed forty-two countries, while the distribution of our meat is following the same pattern.

Chapter 24

MECHANISATION AND ITS LIMITS

In this age of the machine there is a tendency to think everything can be mechanised. Because the grain and crop-growing industries in countries like U.S.A. and Russia can be almost completely mechanised, some assume the same system can be applied here. This is far from the truth. Our material welfare is based on an efficient livestock industry, situated for the most part in broken and hilly land, in contrast to the huge plains and rolling country of the big grain-growing areas of the world. The management of livestock is a highly individual business in the sense that success or failure turns very much on the knowledge, judgement, and industry, of the man in charge. Likewise caring for stock is a human activity in which a man's chief assistants are his horse and dogs, although, on easier land, tractors and wheeled vehicles are used, particularly at lambing time.

This would be an appropriate place to refute the conception of the farm as a 'factory' that is often propounded. A factory is static —so many machines in a building—it is provided with 'input' as the economists say, of so much raw material, power and labour, and the result is a predictable 'out put' of so many finished goods. A farm is quite a different proposition—it is dynamic, dealing with living things, and its output is affected by many unpredictable imponderables. Prominent among these are the weather, stock or crop diseases, soil deficiencies, the quality of the management, and so on, So it is that despite the best management, land and materials, a drought can wreck all estimates of production and cause unforeseen heavy loss, and completely upset all the 'projections' of the economists.

Mechanisation can play but a limited part in our type of farming. But over a comparatively few years we have exploited it wherever it can apply. Twenty-five years ago we cut any logs or timber with a two-handed crosscut saw. Today one man cuts the same log in a fraction of the time with an engine-driven chain saw. We did all our own hay making with horse drawn mower, rake and sweep. Today we cut the hay with a tractor-mower, but the tedder is hired to us by a hay-baling contractor, who then bales it for us with an expensive pick-up baler at a contract rate.

Practically all shearing, in the North Island, at any rate, is done by machine, and contract shearing is replacing other methods—at a considerably increased cost. Today most stock is trucked to and from sale yards and to freezing works, and although it is an added cost, I for one pay it cheerfully, with memories of hot dusty hours dragging along behind a mob of tired sheep, which I have no wish to repeat.

The versatile wheeled tractor, with its built-in hydraulic lifting gear, and the range of implements that can be mounted on it, or drawn by it, is something of a mechanical marvel that the young farmer takes as a matter of course. It has superseded the horse team, as has the tracked type tractor, used for many purposes such as bulldozing and dragging giant discs over rough country. The bulldozer, with a nine or ten foot blade, has been a useful tool which has made thousands of miles of farm roads on hill country.

I recall our first experiences of bulldozing roads, and the fascination of watching a five-ton monster biting into the hillside and pushing the loosened soil out to form the filled side of the road. Relentlessly it advanced on its caterpillar tracks, literally taking everything before it. This exercise of brute strength developed from an efficient diesel engine appealed very much to those like myself who had hitherto relied on the pick, shovel and axe. There was great satisfaction in seeing the machine's contemptuous treatment of large stumps which normally involved a lot of digging and chopping. 'Can you handle this one Charlie?' one would say to the driver. 'Watch this,' says Charlie, as he carefully edges his blade under the flange of a main root and 'hey presto' up it comes and over goes the stump down the hill. We had all kinds of drivers, some of them too venturesome for my liking, who would work their heavy machines dangerously far out on loose soil on a steep slope. Duncan was the driver of an elderly machine that used to develop defects. His universal tool was a heavy hammer used for altering the pins holding the blade. One or two well-placed bashes with the hammer in the right place seemed to be his standard remedy for a stoppage! Few farmers could afford this kind of equipment, but we were very glad to be able to hire these machines at hourly rates. Now these and more modern earth-moving machines are commonplace, including trench and hole diggers. We seldom stop to think what a load of heavy manual labour has been lifted from man's shoulders by their use.

Not long ago we spread fertiliser on the hills by hand as I have related, and on the easy country either by hand or by a horse-drawn distributor. Today nearly all the lime and fertiliser spread on flat or easy land is spread by contractors, operating trucks with various types of spreader mechanism of New Zealand design. But the most dramatic instance of mechanisation, and the one with the greatest possibilities, is the harnessing of aircraft to our pastoral industry as an implement to spread fertiliser, lime and grass-seed. In addition to this aircraft are being used to transport fencing material to difficult hill areas, and to spray noxious weeds.

The most novel use of all is in some of our northern scrub-covered hills that are too rough to be cultivable, where the whole process is done by aircraft. As ground implements cannot be used, the scrub cannot be crushed to dry it out for burning, and cutting by hand is too costly. Desiccant, mixed with Diesel fuel, is sprayed by air on to the standing green scrub about a month before it needs to be burnt. This has the effect of 'browning off' all growth, which will then burn readily. The subsequent operations of liming, fertilising and seeding are all done by aircraft. This technique widens the range of broken hilly country now in scrub and fern that can be converted into productive pastures. Until now such land has been regarded as an almost complete 'write off' having no value, except perhaps for re-afforestation, if a tree species to thrive on this hard soil could be found.

All these mechanical aids save time and labour, and enable men to do more in a working day with less effort, but, most of our pastoral country being so broken, we will always have need of a large labour force, and seldom have enough men. There are for instance tens of thousands of miles of hill fencing on our pastoral land. This was erected and has to be maintained by skilled men. Fencing is almost a lost art, but if we are to maintain existing fences and build all the subdivisional fences that are needed to get the best use of our land, we will need many more men. It is a comfortable theory, for politicians and others, to explain away the reduced man power in our farming industry by saying that mechanisation has displaced them. This is far from the truth.

It is unfortunate that national policy for many years past has resulted in draining men away from the land to protected industries.

These, selling on a high cost local market, and sheltered from competition by tariffs and import controls, have been able to offer inducements that farming, selling on the world market, has been unable to meet.

In a recent survey the Director-General of the D.S.I.R. in New Zealand, Dr. W. M. Hamilton, commented that in twenty-five years, from 1936 to 1961, the male labour force on farms had dropped from 144,500 men to 113,562, a loss of 31,000 men, although in that period the total labour force in New Zealand had increased by 163,000.

He said, 'We have fostered and protected secondary industry at the expense of farming', and then went on to state that a unit of labour employed in farming in New Zealand produced 60 per cent more than a unit employed in secondary industry, when the products of both were valued at world market prices. He also said that the net productivity of a unit of labour in farming in New Zealand was the highest in the world—2·6 times higher than in the United States, and 3·7 times higher than in Britain in the post-war period. In his concluding remarks this authoritative and impartial commentator said that if the aim were to maximise real income per head in New Zealand, the obvious course would be to attract labour and capital into the most efficient industry, farming. 'Slowly, even reluctantly,' said Dr. Hamilton, 'New Zealand is beginning to appreciate that the primary industries of farming, forestry, and fisheries, are still the only industries able to sell on a world market and provide foreign exchange. These industries still continue to provide about 97 per cent of the value of exports, and are likely to continue to do so in the foreseeable future.'

This is an impartial survey by a senior public official who is well qualified to suggest to his countrymen that they should take a close look at the direction in which we are heading. In concluding this piece on mechanisation and its relation with the farm-worker, I hope I have shown a proper appreciation of labour-saving devices, while making it clear that there is, and always will be, the need for a substantial force of skilled men on our pastoral farms. It is a worthwhile employment, full of interest, and calling for versatile talents.

Chapter 25

SCIENCE AND RESEARCH INTO FARM PROBLEMS

In the last forty years or so, since I first took on my bush farm, scientific research has made available to the man on the land a great store of knowledge from new discoveries to help him in his work. Earlier generations of farmers had to find the solution to many of their problems by trial and error, and tended to stick to rule of thumb methods that experience had shown to be sound. In spite of this, they suffered losses, because often the key factor in the situation was then unknown. Today we have many of the answers, though some are still being sought. I can illustrate this by two examples in my own experience. In our warm humid climate our young stock, especially lambs, became heavily infested with worms, the larvae of which are picked up from longish pasture. If not dealt with, these parasites quickly pulled the lamb down in condition, and heavy death rates were common. Although we knew the life cycle of these pests, and that they must be countered by dosing the lamb, we could never completely eliminate losses because none of the medicaments then in use were fully effective.

Today we have practically eliminated loss amongst our lambs from this trouble because a new drug, the result of research, is almost 100 per cent effective against these internal parasites. The only catch in this is that the sale of such products is a tight monopoly held by large chemical combines, who make us pay through the nose for our requirements.

Another example of the 'missing link' relates to trace elements. When we first started topdressing our hills by hand, we used superphosphates, knowing that our land was short on phosphate. The results were disappointing in the extreme, all this heavy labour and expense producing little visible result. It was not until soil research into trace elements revealed the knowledge that a deficiency of molybdenum was the key factor on our type of soil that we began to get worthwhile results. Today we apply molybdic superphosphate from the air, which gives a startling increase in clover growth. The quantity of molybdenum is microscopic, one pound of molybdenum to the ton of super, or 2½ ounces to the acre. This tiny quantity is enough to activate the bacterial colonies on the roots of the clover plants whose function it is to fix nitrogen from the air. This, with the phosphate, results in a strong growth of clover and consequent enrichment of the pasture.

Another example of help from new discoveries is in the use of antibiotics to counter various infections in stock. Every modern shepherd carries in his kit at lambing time a tin containing tubes of penicillin, with a needle for injection, where lambing difficulties are likely to result in septicaemia. New preparations used in sheep dips or sprays enable us to give our sheep better protection against ticks, lice and blowfly strike. Synthetic hormone preparations help us to more effectively control weed infestation by spraying, both from the ground and air. These upset the metabolism of the plant and cause it to 'go hay-wire' and ultimately die.

Another facet of help derived from new scientific knowledge is in the improvement in the processing and qualities of our produce. We have Dairy, Meat, and Wool research institutions, financed partly by the producers concerned and partly by the state. Improved processing, diversification of the product, and endowing it with superior qualities are the main objectives. A striking example is what is being done with wool to improve its competitive position with the man-made fibres that threaten its traditional markets. Various treatments have added to the natural versatile qualities of wool new characteristics that make it the outstanding fibre for most types of apparel. Today the customer has, as a result, the choice of many beautiful and durable fabrics. Other uses for wool are being found, and traditional ones expanded, a good example being the greater use of wool carpets, the best of all floor coverings. New knowledge of genetics is helping us to breed the best types of stock, and to fix good characteristics more certainly and in a shorter time than hitherto.

By means of levies the farmer contributes annually a substantial sum towards research, which is broadly matched by government contributions as an acknowledgement that the general taxpayer also benefits from better farm production. This we regard as bread cast upon the waters, which over the years has given us, as I have related, valuable knowledge. Much of this money goes into fundamental research, which involves long term projects not likely to pay a dividend in practical results for a long time. Some of the money appears to be wasted, in allowing scientists of the type who have little sense of time or money to pursue dreams of their own with rather hazy objectives. This seems to be an unavoidable phase of scientific investigation which has to be permitted in the hope that some side result may be valuable, as it often is. The sums devoted to applied research in general are for the solution of some immediate problem and often come up with a rewarding outcome.

All in all, the 'know-how' that has come out of all the various ways, and there appear to be many, in which scientific work is carried on, has been of great value to our industry. Had it been otherwise we would not have been able to persuade the farmers to continue their contributions. I speak as a farmers' representative who for the last twenty years, as a member of a research organisation, has had a say in voting considerable amounts of the farmers' money for this purpose.

The public conception, 'image', as modern jargon has it, of the farmer, has been in the past of a rather conservative stodgy individual, at least in older lands. It is not true of the New Zealand farmer, whose open-mindedness and willingness to 'try anything once' has made a substantial contribution to our high farming output. When I first attended Lincoln Agricultural College, sixty years ago, the idea of agricultural education was only slowly being recognised, and there were still those who scoffed at the idea of a young fellow 'going to school' to learn farming. All that is changed now, and the farmer is avid for up-to-date information, and quick to act on it.

A clear indication of this is the number that turn up for the well-staged annual farm instruction days put on at Ruakura research centre and at Massey and Lincoln. The officers of the Department of Agriculture do a splendid job in translating the knowledge gained by the scientist into practical use on the farm, and their mana stands high with the farming community. I am sure that farmers generally would approve my paying a high tribute to these men, particularly the field advisory officers, for their devoted work, and would support me in the opinion that, having regard to the effect of their work on production, they deserve better status and pay. They are truly key men in supporting the good life of New Zealand. This Department presents the results of experimental work in its monthly journal in the form of articles. Reprints of most of these are available, either free, or for a shilling or two, at local offices. This is the cheapest technical library I know of. Hundreds of thousands of pounds have been spent on these research projects. There are the results, all neatly printed and illustrated, and available to the young farmer for a few shillings. Fortunate fellow!

Hereford cattle at drafting yards. These are built entirely of split timber.

Treating sheep for footrot by walking them through a formalin bath after paring their feet.

The shepherd—proud of his horse and dog.

The author and friend.

How the kids went to school.

Chapter 26

PLEASURE AND PROFIT IN TREE-PLANTING

New Zealand with its large areas of rain forest, in the North Island particularly, had many species of valuable native trees. As I have related earlier, most of the millable varieties were extracted and used for building, cabinet-making and other purposes prior to the felling of the bush for settlement. Included in these varieties were trees such as the totara, kauri and puriri which provided all the early material for fences, yards and farm buildings. The most generally useful tree was the totara, which was easily split for posts which lasted up to thirty years and more in the ground. Millions of such posts have been the mainstay of our fencing system on farms. Now we have just about exhausted our supplies of these native timbers, and we have to find substitutes for them, both for fencing and building.

Advances in the science of timber preservation, both by simple 'cold soak' methods and the more elaborate pressure treatments, involving the use of suitable chemicals, forced into the timber under pressure, are now playing a big part in this. We are also fortunate that our climate and soil conditions are conducive to the fast growth of these trees, particularly soft-woods such as the Monterey Pine, known in New Zealand as Pinus Radiata. The good timber varieties of the Australian Eucalypts also grow readily with us, far faster than they do in their natural habitat in Australia. The Japanese cedar, Cryptomeria Japonica, one of the best timbers of Japan, also thrives in our environment. Redwoods, oaks and a wide range of exotic trees, including many ornamental ones, grow readily.

Consequent upon all this, farm forestry is assuming a growing importance, both for the supply of fencing and building needs on the farm, and to supplement the output of commercial and state plantations for our future timber requirements. On almost any farm there are areas that can profitably be devoted to tree planting, and there are few things in farming life more satisfying than to see the trees you have planted growing up into valuable timber, or by their beauty enhancing the environment one lives in. As a farm improvement it is not a costly one, and state loans are now available for it under approved conditions. I myself have planted trees extensively, and would encourage any farmer to do the same. It is an activity he will never regret. Sound advice on the species to plant, and cultural methods to produce the best ultimate result, is freely available, and the wise planter will make use of this. Exotic soft woods, notably *pinus radiata*, have a special value because of the readiness with which they will absorb preservatives, such as creosote, in the simplest cold soak method, easily applied on the farm.

The farmer who establishes a properly planned pine plantation, with the optimum spacing of trees, and road access, can expect his first return in ten years time, so rapid is tree growth under our New Zealand conditions of ample rainfall and mild climate. These will be the first thinnings, large enough for fence posts, and easily treated with creosote to give them long life. Pruning of the trees, to suppress branching, and further thinning, will provide both valuable fencing timber, and, ultimately, the maximum of clean timber for milling. Farm forestry is growing steadily and well deserves the support given to it by the state.

Other exotic trees such as the Australian Eucalypts, the Japanese Cedar, Redwoods, and certain varieties of poplar, are well worth planting. Just over thirty years ago, spread over five years, I planted ten thousand eucalypts of good timber varieties, Eugenoides (Tasmanian Stringy Bark) Pilularis (black butt) Saligna and Botriodes, of which Saligna has been outstanding.

We have set up our own farm sawmill and are now milling the final thinnings, leaving the dominant trees widely spaced to mature as good milling timber. These are expected to reach 3 feet in diameter at the base in a further ten to twelve years. Many are already over 130 feet high, 80 or 90 feet to the first branch, giving a long clean bole. They delight my eye when I walk through the roads we have formed through the plantation, and I find it hard to believe these have grown from the seedlings 8–9 inches high I planted amongst the fern.

The thinnings we are milling average 18–24 inches in diameter, and provide strong densely grained timber for gates, rails, for yards, and fence battens, despite some tendency to warp and split. A feature of the eucalypts is the high proportion of heart wood, about 80 per cent, which is in marked contrast to many of our native timbers. The wood of these eucalypts is too dense to take preservatives readily, but the heart wood appears to be durable above ground. The establishment of plantations fits well into the farm programme as the fencing and planting, and later the thinning and pruning, can be done in the winter months when the pressure of the seasonal work has lightened. The cost is not high, especially if the work can be done by those normally employed, and if the fencing posts can be produced from homegrown timber. It is important to choose varieties that will thrive and be happy in the particular type of soil and climate. Some varieties of eucalypts do not like a clay subsoil—others, such as the saligna, one of the best, do well on it. Some trees like the seaside, some do not, and they vary in their tolerance of frost.

An aspect of reafforestation that deserves more emphasis is the propagation of some of our most desirable native species. This is being done at some of the State Forestry stations, notably at the Waipoua Forest, where much work has been done on regeneration of the kauri. This is very long term work. The native tree that to my mind is outstanding in suitability for replanting is the totara. On all northern hill country we have had good value from the handsome and often mature totaras that were scattered through the original rain forest. Many of these were forest giants four and five feet in diameter with long clean boles. They provided us with millions of fence posts and much top grade milled timber, especially that used for joinery.

When the bush was cleared and burnt the green totaras were usually left standing. The fire killed most of these trees, which were subsequently felled and either split for fencing or hauled out for milling. After the land had been sown to grass for some years, seedlings, from seed that had lain dormant in the ground from the time it was shed by the original forest trees, began to appear. Sheltered by the remains of the burnt bush, resilient and flexible, and protected by a spiky foliage, many thousands of these young trees grew up on hill bush clearings. This is the one and only valuable tree known to me that can effectively protect itself from damage by stock. It is used as a hairbrush by the bull, and is played with by goats and sheep, but up it bobs on its tough flexible stem and goes on growing. This seems to give it a special value for planting which has not been sufficiently appreciated.

On this farm we now have an estimated number of 5–6,000 of them, many already of fair size, 30 feet high, mostly in groves. They are splendid shelter for stock, and where they come up too thickly the thinnings when trimmed and pressure-treated make good posts. Because the stock manure fertilises them, these trees grow much faster than the forest trees, and are not the long-term proposition that kauris are. With these totaras, and eucalypts, jap cedars, and pines I have planted, I estimate that the farm should be self-supporting in all its timber and fencing requirements for the next 100 years. This provision could be made on many farms. In a timber-hungry era that seems increasingly likely to eventuate in years ahead, I can hear one of my five grandsons say to another—'Well, yes no doubt the old boy did have some cranky notions, but you've got to admit he had the right idea about trees!'

I cannot close this chapter without a piece on the planting of ornamental trees. The selection and spacing of ornamental trees as single specimens, in driveways, or plantations, can transform the appearance of a farm homestead and create surroundings of beauty. Not only does this please aesthetically, but in practical terms it enhances the value of a farm. Many lovely settings for farm homes can be seen in some of the early settled large farms of New Zealand, where the owner was evidently a tree-lover with memories of the handsome trees of England or other lands. It is an ambition of mine to encourage today's generation of farmers to emulate them. I especially commend this idea of planting ornamental trees to elderly farmers, like myself, who have turned over the executive management of their farms to sons or managers, and have time to devote to this kind of thing.

Young men are inclined sometimes to be a little scornful about ornamental plantings, preferring to concentrate their activities on plantations of more commercial value. I suggest to those getting on in years that one of the best things we can leave behind us is a legacy of beautiful surroundings in the environment where we have spent the active years of our lives. Small plantations of selected trees, in driveways, or the approaches to a farm homestead, can create a charming effect which can delight visitors as well as the occupants of the farm. Such areas must of course be fenced off from stock, but this is not a costly or formidable proposition. If the 'old man' will undertake the planting and care of the trees, the younger generation can usually be prevailed on to help with this. Spotting individual specimen trees, chosen for their beauty, around the home paddocks takes a bit more work, as each tree must be individually protected from stock, especially horses, who are very destructive and seem to possess telescopic necks. A triangular construction of three posts spaced 6–7 feet apart, with rails between them and three feet of netting above, gives good protection to single trees and is proof against all stock.

Nurserymen are always ready to give advice on varieties that can be expected to thrive in the locality, and it is important to study this, as it is frustrating to lavish care on a tree only to find it fail. As suggestions, I mention here some of the trees which I have found to do well, though they may not be happy everywhere. High on the list is the Liquid Amber, easily transplanted and a strong grower, with ornamental foliage providing lovely autumn colours. The scarlet oak, *Quercus Palustris*, is another handsome tree with an autumn display, and the golden elm with its crown of yellow golden leaves standing out in sharp contrast to the green tones. This requires a dampish strong soil. Then there is the golden ash and the claret ash, and the melia with its drooping foliage and pendant blue flowers.

The flowering horse-chestnut with their cream or pink candles make good specimen trees, as does the paulownia. Golden willows and poplars can give attractive autumn colours in damp spots. The deodar makes a handsome and quick growing specimen. One could also mention the gingko and the fossil tree and many others. I do not wish to make this a technical treatise, since there are first-rate books to be had on the subject, but I do hope that what I have written may interest my readers and encourage some of them to plant trees, which will give them enduring satisfaction and pleasure.

A delightful dividend from tree-planting can be the increase in bird life. Apart from the usual sparrows, blackbirds, thrushes and larks, we have on this farm lots of fantails, rosella paroquets, kingfishers and white-eyes. The Maoris call the last mentioned 'little strangers' as they are self introduced from Australia, no doubt having been blown across in westerly gales. The 'fannies' with their darting flight are fascinating, while the paroquets and kingfishers add flashes of vivid colour to please the eye. On one occasion I heard a kookaburra and located him perched, appropriately enough, on a high branch of a Sydney bluegum. He must have been a visitor from the colony at Kawau Island having a look round, but evidently decided not to bring his family, for I have never seen one since.

In New Zealand we have relied on *pinus radiata* for exotic timber because it grows so fast and takes the various preservative treatments so readily. I rather doubt the wisdom of carrying on with this monoculture of only one species when overseas experience, particularly in Germany, has shown the value of mixed planting of varieties that thrive in association. Is our Forestry Research Institute at Rotorua making any study of this?

Modern methods of tree culture in spacing, thinning, and pruning, are directed to ensuring a higher yield of clear millable timber from each tree, but there is still a substantial wastage in conversion to usable products, mainly timber and various types of wall-boards. I visualise the time when much that is now discarded as cut-over trash will be utilised. The whole tree, including its trimmed branches, will be sent to the revised version of a timber mill, where everything will be minced to small cubes by multiple fine saws. These will go into a chemical digester, to be reduced to a semi fluid pulp, to which will be added materials to make the final product both fireproof and vermin proof. Then the mass will be run out into moulds which will shape girders, scantling, flat sheets and other usable shapes for prefrabricated buildings.

Chapter 27

FARMING AS A CAREER— WE NEED THE BRIGHT BOYS

I am sometimes asked to advise on the possibilities in farming as a career. The typical case is that of the business or professional man whose son wants to be a farmer. The father's difficulty is that he sees no prospect of finding the necessary capital to give his son a start. There is no doubt that many potentially good farmers have been lost to the land for this reason. In the national interest we must alter this, and there are now some signs of a change. New Zealand's prosperity is likely to be closely based on her export of primary produce for many years to come, so that the knowledge, skill, and industry of her farmers is vitally important. In a vocation calling for the widest knowledge, we should be recruiting our brightest young men for farming.

If we can train and successfully settle on the land approximately 13,000 ex-servicemen of World War II, many of whom had had no previous farm training, then surely we should be doing the same for the on-coming generation of potential farmers. The immediate economic future of New Zealand lies with only a few thousand of these key young men, and we would do well to recognise it and act accordingly. The farm workers of today, whether sons of farmers or employed men, form the reservoir of skill and knowledge from which the farmers of the future must be drawn. It is a tragic mis-use of talent when, as so often happens, such men take town employment because they see little hope of acquiring a farm of their own. This should not be.

What is needed is a properly organised training scheme, backed by the State, in which selected young men would be apprenticed to approved farmers. During their training period they would attend special short courses at one of the agricultural colleges. On completion of the apprenticeship period they would be required to take a job as a farm worker for some minimum period. Today this is a well paid job, and if our trainee wants to marry he will have little difficulty in getting a position which will provide him with a reasonably good house and perquisites. Throughout this time his progress and capabilities will be periodically checked by a small local committee, as was done with the ex-servicemen, and he will be graded as to capability. When he reaches the top grading, and on completion of a total minimum period of training and as a farm worker, he will be guaranteed an opportunity of acquiring a farm of his own, as was done with ex-servicemen. He will be expected to save during his qualifying work, and some special exemption from income tax should apply to the money he is able to put into a savings account. He will be expected to make a 'down payment' of an appropriate amount on his farm, the State financing the rest on a long term amortised mortgage at concessional interest rates.

There is nothing impossible about all this, and some of it is already being done. The Government has recently announced that it will subsidise a training scheme inaugurated by Federated Farmers, while the State advances corporation is now greatly expanding its loans for farm finance at attractive rates of interest below those of commercial lenders. This is good as far as it goes, but the whole thing needs to be pulled together into a well-organised scheme with State backing, and a guarantee to the young man who enters it, that when qualified on the lines suggested above he will have the chance of farm ownership. There would be little trouble in finding experienced men with the knowledge, sympathy and appreciation of its importance to run such a scheme. Without such a plan, and failing some responsibility on the part of the State to ultimately provide farms for these young men, the high capital cost of getting a young farmer started must act as a deterrent, and lose the industry and the country many desirable farmers.

The total capitalisation required to start a young man as a sheep farmer is roughly £20,000, of which he requires about a third as his own capital. For dairying it is lower. With the State assistance I have suggested, the individual's personal capital can be quite a bit less, though he would still require a substantial sum. Over the period he is qualifying it means some self-denial and plenty of hard work, especially if he undertakes outside contract work, such as fencing, in holiday periods, as some do. On the other hand, farm wages today are higher than they have ever been, the catch being what has to be paid in income tax. A good case can be made for exempting from tax whatever amount can be paid into a special savings or trust account.

The qualifying period, and the need for self denial in accumulating capital, is likely to choke off the faint-hearted. It will not deter the boy with a real love of the land and stock. I can assure him that to become a farmer in his own right is an ambition worth some struggle and sacrifice, and one which will give him a satisfying life equalled by few others in interest and the fullest development of personality.

My advice to the business or professional man with a son keen on farming is—give him his head. Try and give him a course at one of the agricultural colleges, which will open a wonderful world to him. Let him take a farm job. Competent men are scarce and can command a good wage today. Even if he ultimately does not go on the land, the wide all round experience he will get will still be of use to him whatever he does, and the effect on his character and health can only be beneficial. It is a shame to deny any boy keen on farming the opportunity of such an interesting and fulfilling life, and his chances of getting a farm of his own with State assistance seem to me to be substantial. The country *needs* such young men, and will be compelled to give him his opportunity sooner or later.

Purely as a money investment farming does not rank high. But as an investment in other ways it has much attraction, and this is how it should be regarded when considered as a career. Let us have a look at some of these other values that too often are overlooked. Farming is a way of life, where one lives close to nature with all its interest and beneficent effect on both mind and body. It is a physically healthy occupation, in which one's body is exposed to the sun, wind and rain it was designed for, and does the physical work that keeps it fit and well. Mentally there is a constant challenge—such a wide range of knowledge is needed, so many interesting things turn up day by day, and one has to apply one's mind to such a diversity of problems. Consider some of the subjects on which the farmer needs to be informed.

It all starts with the soil, a complex subject in itself. To get the best results he must understand something of the formation and composition of the soil types of his land. When he has this taped, he will know how best to invest his money in fertilisers to meet any deficiencies in both main and trace elements. He has to know about different grasses and the tricky business of pasture control by stock. He must have a good knowledge of sheep, cattle, dogs, and horses, and know how to keep them happy and thriving while still using them as implements to regulate the growth of pastures. He must know the points of a good animal and what to aim for in what he sells, and as a breeder he must have a working knowledge of genetics.

He must be a fairly good amateur vet, and able to act in the capacity of midwife to large numbers of ewes and cows. Some idea of chemistry is necessary in an age where we use a wider variety of stock remedies, weed hormones, dips, and sprays, than ever before. He must be a carpenter and mechanic of some ability, and if in the real blackblocks he should be something of a blacksmith and farrier, able to shoe a horse. The provision of water supply, for house and stock use, calls for some competence as a plumber and drain-layer. An understanding of markets and when and where to sell his products, such as stock and wool, is necessary. He must be a good fencer, able to erect any type of fence, build and repair yards and hang gates.

He cannot afford to be ignorant of farm financing and everything to do with advances for development work, and finally he needs to have some ability as an accountant with a working knowledge of the deplorably complex system of taxation with which we are saddled. Versatile?—Of course—and calling for endless acquisition of information . . . which makes farming life so interesting. Then there is the farm as an environment and background for the family as touched on before. There is none better, with the open air life and association with animals and nature.

There is the effect on character. The man on the land works with nature, he reaps what he sows, and is successful to the extent that he conforms to natural laws. He does not get away with anything, and pays for his mistakes. He meets with 'Triumph and Disaster' and learns to take them in his stride. He develops an acute awareness, and learns to look ahead. Because he is usually short-handed, he makes the best use of time, and is one of the world's best improvisers. All these factors in his environment inevitably mould his character, and make of him 'a man of many parts'.

The farmer's son who ultimately takes over from the 'old man' is a big factor in our farming whose way is often more difficult than it should be. Creating and building up a farm is a very personal matter, which begets a strong possessive feeling in those who have done it. Many find it hard to relinquish this as age overtakes them. The wise farmer fortunate enough to have one or more sons keen to follow in his footsteps, tries to see that, when the time comes, the transition of authority and ownership to a son or sons is made unselfishly and fairly. He should try to give his son the advantage of training at one of our Agricultural Colleges in addition to experience on other besides his own farm. He should pay him full wages when working at home and not regard him as 'cheap labour'.

As the young man gains experience means must be devised to give him the opportunity of acquiring an interest in the property. At this point it is as well to pay for the advice of a sound solicitor and accountant. So absurd is our tax system, that the farmer who has built up a highly productive unit often cannot afford to go on farming it as an individual, so penalised is he by tax rates on a graduated scale. Formation of a farm company, a trust, or partnership, usually has to be considered, which offers the opportunity to spread ownership in the family. This also offers a means of mitigating the effect of harsh death duties. There have been far too many cases where the estate of a farmer has been so crippled by taxation upon his death, that the son, or wife, who has to carry the farm on, is left with an unfairly burdened proposition.

If a farmer is still fit and well in his old age, it is often hard to make up his mind to retire, but he must consider the young who are to follow him, and at the right time hand over executive management of the farm to his son, subdivide his farm or do whatever is appropriate in the circumstances. All this is not to say that he should not retain sufficient financial interest in the property to provide for his own old age, and ensure the position of others who will have to be provided for from his estate. If the 'old man' elects to continue living on the farm, he can find plenty to do, but, having handed over the management, he must resist the temptation to 'butt in', and give the younger generation a chance to work out their own ideas. This takes quite a bit of self-discipline!

The idea that upon retirement the farmer should remove himself to town is in my view a poor one. Too often he finds himself in a foreign environment in which he has little interest or scope for the physical activity he needs, and he is bored to tears. Far better for him is it to stay where he will have plenty of interest and useful light physical work in surroundings he understands.

In closing this chapter I commend to any reader who can exercise any influence with the Government of the day, the idea of a state-sponsored farm training scheme with a guarantee of eventual settlement for the trainee. This should include the suggested tax concession on what he can save from his earnings while gaining experience. Various governments have dodged the issue on this matter. It is now becoming essential, and in the present climate of 'more production' would have a good chance of acceptance if resolutely pushed. Let us build up an élite from the young men working on farms today who will ably carry on our industry in the future. Could New Zealand have any better investment? As Churchill used to mark his minutes, 'Action today please'.

The reader who does not live in New Zealand will realise that all the above is in the New Zealand context, and relates only to conditions in this country. This is deliberate, and is intended to give a fair picture for the information of anyone from overseas contemplating becoming a farmer in New Zealand. I shall elaborate a bit further on prospects for the newcomer in a later chapter.

Chapter 28

REFLECTIONS ON OUR NEW ZEALAND WAY OF LIVING

Having tried to give a picture of farm life as I have lived it, which I hope is a representative picture of many other farms in the North Island of New Zealand, may I now offer some observations as a responsible citizen of the land where I was born seventy-five years ago. We have developed a strong democratic tradition in New Zealand and with it have come as close to a really democratic way of life as any country. We believe, and act on the principle, that, in our own idiom, every individual should have a 'fair pop'. That is, that justice and fair dealing should prevail in practice as well as theory, and that everyone should have a reasonable minimum standard of life and the pleasures and opportunities that go with it. We think that our national production of wealth being what it is, that no one should endure real poverty and its miseries.

To that end we have developed what has come to be known as the 'welfare state'. That we have been able to do so, and carry it as far as it has gone, is in no small measure due to our high export earnings. It is time we stopped to consider just how far we should go with direct benefits provided by the State at the cost of the taxpayer. In a two-party political set-up, such as ours, there is a strong temptation for either the 'ins' to enlarge hand-outs to maintain themselves in power, or for the 'outs' to bid for votes by promising certain sections more from the general pool. Figures show an alarming growth of expenditure on social services. This, if it goes on, can defeat its own ends by retarding or choking off the main productive industries that must provide the bulk of the taxation revenue. We are already one of the heaviest taxed countries on earth, and there is not any doubt that the weight of taxation has a restrictive effect on the output of farming and other industries.

There are other results more serious—I refer to the damaging effect on moral fibre and national character. Initiative, industry, and other pioneer qualities, have transformed New Zealand in little over a hundred years from a virgin land to our present state. The steadily growing idea that the state should provide more and more, is undermining and eroding the very qualities that made our country what she is. I suggest that it is time some study was made of all this, and that there should be some sensible middle course.

We would all agree, I think, that education should be free or nearly so, that the sick and aged should be cared for, and the incapacitated and widows provided with the means to live, and that other similar responsibilities should be met by the taxpayer. On the other hand the State is handing out millions in directions that could be pruned without real hardship to anyone. Food and milk subsidies, for example, seem an absurdity in a country with average earnings as high as ours.* Since no government seems likely to have enough guts, unaided, to check this growth properly, could the problem not be examined by a Royal Commission? Such a body, composed of eminent men representing all shades of political colour, would perhaps be able to make recommendations that would be generally acceptable, or that a strong government would act on.

New Zealanders are highly literate, being amongst the highest *per capita* buyers of books anywhere, and we travel a good deal both within our own country and overseas. Nevertheless we are a bit touchy about criticism from visitors, and tend to be a bit insular. Sometimes the criticism is valid, especially when it refers to such things as the restriction of initiative and a 'couldn't care less attitude' induced by over-reliance on State benefits. Conversely I think we are entitled to question the validity of such comments when they come from prominent people that think the 'push-button' civilisation of the highly industrial countries is the ultimate in civilised living. Is it? I for one very much question it. Is it a good thing to have millions of people living in the unnatural conditions of crowded urban areas, many of them in multi-storey buildings of flats, even if amply provided with television sets and every modern gadget?

Is it not perhaps better to have our people more spread out, owning homes and gardens of their own, with good access to beaches and places of beauty, and the opportunity for the young to participate in, rather than merely watch, all kinds of healthy sport? This is our New Zealand idea, and we should count our blessing and not undervalue this kind of thing. What finer demonstration of the democratic way of life is there than the thousands of people to be seen on sea beaches in holiday periods. Here we have Tom, Dick and Harry—

* Some of these subsidies have now been eliminated (1968).

the truck driver, the carpenter, the accountant, the doctor, and the farmer, pakeha and Maori, all having the time of their lives with their wives and families. The kids have a marvellous time bathing and building sand castles, while father fishes or bathes, and mother bathes with her brood or relaxes on the beach.

All this on beaches everywhere over the country that are free to anyone. There are beauty spots, skiing slopes and lakes easily reached, each offering something that in many other lands would be the monopoly of the well-to-do.

A really typical New Zealand occasion can be seen during the Toheroa season on our West Coast northern beaches. On a fine day, at the right state of the tide, hundreds of people from every walk of life, complete with kids, prams and all the family doings, can be seen scrabbling for toheroas for miles along the beach. The more sophisticated and venturesome take their cars on the beach, and when they get stuck in the sand, as some do, all hands rally round to push them out. It is a marvellous day's outing for a family, with a fine sandy beach disappearing from sight to the north, and the huge ocean combers crashing into surf in the sunshine. The spice of discovery of a good patch of shellfish adds to the interest, for they are not to be had for the mere picking up. The tiny circular markings that denote their presence have to be sought for and recognised and much vain digging goes on as the less experienced furiously excavate in the wrong places.

But it is all good fun, and as the family returns to its car or truck laden with their full quota of 100 per car there is eager anticipation of the delectable soup or fritters soon to be enjoyed. These shellfish with their unique flavour are a delicacy found nowhere else. The plankton which nourish them gives the waves a brownish tinge as they curl over to break on the beach, in which the tohera are to be found at depths of about 6–9 inches in the firm sand at about half tide level.

The seagulls have their own method of feeding on these shellfish. Seizing them in their beaks, they fly up to 30–40 feet and drop them on hard sand or rocks, which breaks them open. They follow the digging parties with interest and are quick to grab any undersized toheroas left on the surface. None under 3 inches in length may legally be taken.

We have much to be thankful for—the opportunities mentioned above, the natural beauty of our land, a climate without extremes, and the friendliness of people which so appeals to visitors. Let us see that we perpetuate these things. Periodically we hear moans from professional people, usually those whose salaries are provided by the taxpayer, that they could earn more overseas. As a country of only two and a half million people we will always find it difficult to compete in salaries with countries having greater tax revenue to draw on. But in living, and bringing up a family, considerations other than money must count. Too often such people fail to value properly our New Zealand environment with all it has to offer, especially to children. Quite a few return to New Zealand after experience in other lands.

It has often been said that New Zealand has led the world in some of its experimental social legislation, and I offer for consideration, of the reader a further field in which we might give a lead to older lands. Despite having one man one vote, which is regarded (too readily I think) as the ultimate in democratic self-government, we do a good deal of moaning about our governments as one succeeds another. In practice we find ourselves, both here and in Britain, committed to a two party system with little alternative offered to the voter. This makes for weak government, whatever party may be in power. The governing party has a constant eye on the 'floating vote'—the minority of uncommitted electors that determines the result of an election. This means that it overspends the funds the taxpayer provides in directions designed to make itself popular, and is too often reluctant to take the unpopular measures that good government demands. In practical terms this means that we are permanently committed to continuing heavy government expenditure in all sorts of projects, with its concomitant of heavy taxation and continuing inflation. Any government that had the 'intestinal fortitude' to take the strong measures that would be needed to counter inflation and make significant reductions in taxation, would not survive the next election. So it goes on, and we stagger along, with booming values in all types of real estate, continuing depreciation in the value of all savings, and in the purchasing power of fixed incomes, and a succession of upward revision of wages and salaries. The worst feature is the insidious and steady rise in costs which adversely affects the exporting industries, the main foundation of our economy.

Every individual should have a say in the appointment of those who are given the power to rule us, and one must support the principle of everyone having a vote. But I do not think this should be regarded as the best that can be done. I suggest we should examine the idea of the qualified vote, under which those who undertake community responsibilities above the average—not relating to their possession of money or property—should have one or more extra votes. Does it make sense that an irresponsible youngster, because he or she has reached the age of twenty-one, should have the same voting power as, say, a judge of the Supreme Court?

What about those who unselfishly give time and energy to the administration of the many boards, councils and local bodies that are necessary in our complex civilisation? Would they not be worthy of an extra vote, and would they not exercise it sensibly? What about the responsible family man and his wife, who have made a success of their married life and reared two or more children. Would they not be worthy of an additional vote? Then there are those who have been awarded honours for conspicuous service to the community—an extra vote could be considered part of their reward. This idea could start in a small way, restricted to those likely to gain public acceptance, and be extended on the results of experience and public willingness to accept the idea. I commend consideration of this to the reader as an innovation that could lead to sounder government.

In his book *In the Wet* the late author Nevil Shute develops this idea imaginatively. In his story the West Australian government adopts the qualified vote with outstanding results. So effective is it that the Federal government soon follows suit, and Australia then leads the Commonwealth in good government, while the U.K. flounders into depressed conditions under the system that rules today. Might not he prove to be prophetic?

* William Heinemann Ltd.

Chapter 29

SHORT COURSE FOR THE PROSPECTIVE KIWI

I hope that some of those who read this book may eventually become citizens of New Zealand, and that in doing so they will find a fuller life for themselves and their children. I shall, in this chapter, try to paint a fair picture of what the newcomer from Britain can expect, with some information that may be helpful.

Population. At the time of writing, mid-1968, there are just over 2,700,000 of us in a country roughly comparable to the U.K. in area. We badly need more people, and would sooner welcome them from 'the old country' than anywhere else. Some of our facilities and amenities are remarkably good for so small a nation, notably our roads, electric power installations and reticulation (nearly everyone has power) and our standards of housing and public buildings. On the other hand those coming from a densely populated area will miss the frequency of public transport services, and some other things that are more readily available where there are more people to be catered for, and more hands to help. The vast majority of the population is of British origin, the only other ethnic groups of any size being the Maoris and some Dutch. The Dutch, many thousands in number, were post-war immigrants, and have made very good citizens. The new arrival will very quickly find himself at home in a British country.

Climate. This is rated as temperate in the South to sub-tropical in the far north. Rainfall varies from about 20–25 inches in the drier inland areas of the South Island, to 45–65 inches in the sub-tropical North. Rainfall is well spread through the year and a very wide range of plants and trees can be grown. New Zealand is a gardener's paradise, and we are a gardening people, as evidenced by thousands of well-tended gardens to be seen everywhere, many of them very beautiful. I think it would be a fair statement to say that the climate is generally better than that of the U.K. Although we do get the occasional snow in the south, we do not get the prolonged bitter winter weather that seems so often to be the experience on the other side of the world. No doubt the fact that the nearest other land is 1,200 miles away and that we are completely surrounded by sea has something to do with this.

Topography and scenery. Of this it could be said: 'You name it —we have it'. The South Island is lavishly equipped with mountain ranges of great beauty and splendour running up to 12,000 feet in height and providing every kind of alpine sport. Amongst these are some beautiful land-locked lakes. In contrast are the extensive Canterbury plains on which are to be found most of our arable farms; these provide the grain for our bread and beer, and fatten many lambs for export. The North Island is mostly hilly, two snow-clad extinct volcanoes providing ample scope for winter sport. Most of the land was originally clothed in rain-forest as described elsewhere, and now carries millions of sheep and cattle.

Splendid sea beaches all round the coast are easily accessible by good roads, and we flock to them in our thousands in the summer for boating, surfing, and bathing. Motels and camping places cater for the motorist, who, in New Zealand can have a millionaires' holiday taking in some of the finest scenery in the world at very moderate cost. While doing just this recently, I met an English couple who had arrived by ship at Auckland, where they picked up a hire car. In three weeks they expected to see most of the sights, and on returning to Auckland would fly home. Not everyone can afford this sort of thing, but I think they felt they had had good value, and were obviously pleased with their friendly reception. New Zealanders are very 'matey' and do not hesitate to bowl up to you and introduce themselves. When they find you are a visitor, they will do anything they can to help you along and see that you will have a pleasant experience while here.

Industry. Live stock farming is our principal business, divided into dairying as one speciality, and sheep and beef cattle raising as the other. These earn some 93 per cent of the overseas exchange which pay for our imports from the rest of the world. Of growing importance are industries based on the extensive planting of exotic forests, mainly pine, made many years ago. These put out paper, pulp and wall boards and logs for export to Japan. Then there are woollen mills processing our home-grown wool, and a large range of other processing industries preparing the products of our farms for sale overseas. The most important of these is the freezing industry, converting slaughtered livestock into not only carcass meat, but a wide range of pre-packaged meat in cuts, tins and 'freeze-dried' forms. At the moment these meat products are our largest single export item.

An infant industry now being established is our first iron and steel enterprise, based on the smelting of the millions of tons of ironsands on the West Coast of the Auckland Province. We are assured that the technical processes being used are sound, and that the output should ultimately become of economic importance in a country that, to date has had to import all its requirements of iron and steel. There is a wide range of manufacturing industries catering for our own internal market, putting out such goods as clothing, footwear, electrical appliances, textiles, machinery, paint, tyres, car components and what have you. These are small scale industries operating under the protection of import control and tariffs, consequently their products tend to be higher in price than the imports they displace. Another activity is that of car, truck, and tractor assembly on items brought into the country in 'knocked down' form.

Development work of all kinds natural to a young country absorbs a sizeable slice of our national effort. This ranges from huge hydro-electric schemes to supply the ever-growing demand for power, to large public buildings such as hospitals, schools and universities. There is an active policy of state house-building, while the forming and sealing of new roads and improvement of existing ones is constantly in progress. State expenditure on such works, largely financed by loans, is heavy, and there are those who think that for a small country we are trying to do too much too fast. Parallel with the State, private enterprise makes a substantial annual investment in new buildings and facilities.

Employment. From what has been said above it should be clear that whatever a man's background, if he is willing to work he should be able to find a congenial job. At the moment we have what is styled as an 'economic recession' due to over-expenditure by the Government and a reduction in our overseas earnings due to a sharp fall in the price of wool. About $\frac{1}{2}$ per cent of the work force is unemployed, most of them being unskilled workers. Government policy for many years has been one of 'full employment' and every effort is being made to see that no willing worker remains idle. Newcomers from industrial centres in the U.K. looking for factory work will find one sharp difference. In New Zealand there is not the same amount of narrow specialisation on one job that is so often seen in works in Britain. We expect a man to be something of an 'all-rounder' willing to turn his hand to anything that needs doing. If he has not done it before, provided he is not afraid to learn, we'll very soon teach him.

It is the same on our farms—we have not much use for the man who just wants to drive a tractor and nothing else, or the man who wants to do stock work exclusively. On few farms can men be employed solely on such work. For a given number of stock, we employ far fewer hands than would be the case on a farm in England, so employees have to be more versatile. Our men are expected to help with all the farm operations throughout the year. It makes for a more varied and interesting life. They are well paid, and usually have good housing and quite a few perquisites. The skilled farm worker can have a very good life in New Zealand today.

Those who have been used to life in the densely populated industrial areas of England, with mobs of people around them all the time, and mass organised entertainment, often do not take kindly to our New Zealand environment. This is a pastoral rather than an industrial country, and those who like the close living conditions pictured in television in 'Coronation Street' would be well advised to remain where they are. The cosy small local 'pub' with its darts and social atmosphere at the end of the street is not frequently encountered, though the newcomer need not do without his television. Wages will be found to compare favourably with other countries, and minimum rates are protected by awards of the arbitration court. The competent worker is often paid above award rates.

Education. This is free and of a high standard, and the child learns under good conditions in modern and well-appointed schools. Subject to passing a university entrance exam, the youngster from any home has the opportunity of higher education in well equipped university buildings. He is helped by various allowances and bursaries.

Medicine and Hospitals. Under a state run system the patient may visit the doctor of his choice, but pays rather less than half of the doctor's fee. Hospital treatment, including operations, is provided free by the Government, and is of high standard. In non-urgent cases there is often a waiting list.

Other items in the Welfare State New Zealand Style. New Zealand led the world in the institution of old age pensions many years ago.

Since then the idea has blossomed into a comprehensive system of benefits for those in need. The philosophy underlying this has been that everyone should have at least a minimum standard of living, and that those who through old age, illness or for other reasons which are no fault of their own, have lost their income, should be helped out by the community. In addition to old age, widows, orphans and invalids benefits, there is payment for the unemployed, and a family or children's allowance which is at present 15s. weekly for each child. There are various other grants for special purposes. To pay for all this, taxation has to be heavy. Income tax is graduated and there is liberal provision to help the family man. A couple with children on a low to moderate income get off comparatively lightly.

Of interest to intending newcomers are the reciprocal arrangements negotiated with both the Australian and U.K. Governments, which apply to the above. To qualify in New Zealand one must intend to reside here permanently, and a male applicant for age benefit must be sixty-five years or over. Fuller information, including the categories covered by assisted immigration, can be obtained from the High Commissioner for New Zealand at New Zealand House in London.

The people we like to welcome are those who come here in search of a fuller life, and who are prepared to make their contribution, in effort and responsibility, to the building up of a vigorous young nation. Such people will find a warm welcome and a wonderful future for their children. This is the best country on earth for kids.

Chapter 30

TOWARDS A PHILOSOPHY OF LIFE

The biographies of others, particularly those who have had unusual opportunities of observing the world and humanity, have always interested me, in particular the conclusions they have reached about life. I think there is a responsibility on the old to pass on to younger generations anything that may be helpful. Looking back, how often have I said to myself 'Why didn't someone tell me?' A friend of mine said of his father: 'A tough man and wise; but I wish he had passed on a little of his hard-earned wisdom to me. He could, in no time at all, have told me of things it took me fifty years to find out for myself.' We all need some basic guide-lines or principles by which to regulate our conduct.

Education and religion are supposed to help us in this, but what we get from these sources depend upon the individual and how the precepts he absorbs stand up to the batterings of his own life. The impact of what we are taught in youth becomes modified by experience, and when we become confused and frustrated, we tend to doubt what is sound and good as well as that which experience has proved to be illusory. One does not want to be led into 'throwing out the baby with the bathwater'. Although I have become disenchanted with some of the claims of orthodox religion, I think that the Christian ethic of honesty, truth, and regard for others offer us the best set of standards to live by. The old virtues as set forth in the Sermon on the Mount, still hold good. If we base our difficult decisions on these principles we won't go far wrong, whatever the immediate rumpus it may cause.

We have to beware of ready-made conclusions, remembering that so much of the mass media of information today in printed matter, television and radio, is 'slanted' to give a certain viewpoint. The public relations man and the propagandist are loud in the land, and are employed to influence us to view favourably the activities of those who hire them. One of our sages said, truly, 'Man acts not on the facts, *but on what he believes to be the facts.*' This is the basis of all propaganda, and we need to beware of 'baloney' from even reputedly authoritative sources. It is hard to get the truth; we must think for ourselves.

As the years pass you come to realise that many things are not what they seem, or what you have been led to believe they are. In brief, we have to shed many illusions. It rocks most people to find, for instance, that something they have believed in for fifty years is a fraud. This is part of the education of living. It tends to make many cynical, but bitterness helps nobody, least of all oneself.

As increasing knowledge exposes men, things, and situations, for what they are, life can be faced with surer judgement and confidence—that of the realist. As old age approaches one is able to regard the daily scene with more detachment and less liability to worry—surely something to be thankful for. I find myself adopting an air of detachment and some inconsequence, which leads to interest in all sorts of objects not previously seriously regarded. This, it seems to me, should provide a full life for the time that remains to one.

Now to mention some of the things that seem to me important after seventy-five years of life. On gaining experience for example. One does not need to find everything out the hard way, by trial and error. The resistance of a brick wall can be assessed without having to run into it head-on. The wise make use of the experience of others, with one important qualification, that is, to make allowance for the differing circumstances, and to apply the conclusions with judgement. On attaining one's ambition: It is as well to be sure that the ambition is a worthy one. What a disappointment after years of effort to find the hoped for reward an empty one!

A valuable quality in working towards an objective is that of intelligent persistence, or well-informed resolution. We sometimes attribute too much to 'fate' . . . 'He was thwarted by fate, etc.' Chance undeniably plays its part in our lives, but I believe that the resolute individual, by taking the right thought and action, can control his or her circumstances to a greater extent than is generally realised. We talk of the 'lucky' man—isn't he often just a little better informed and quicker off the mark than most?

The world does not owe us a living, contrary to some of the illusions fostered by the Welfare State, and there is no substitute for personal effort. Aldous Huxley says that the essential virtues are love and awareness. Liddell Hart rates accuracy as the essential virtue—to be sure of your facts and not spread misinformation that may lead to trouble. Two of the most profitless human emotions are envy and self-pity. Envy gets you nowhere—the subject of it usually has his own troubles which may not be visible. Similarly, 'that most contemptible of emotions, self pity' as one writer described it, is a poor one to indulge in. In a time of depression which is all part of human experience, we should try in modern jargon to 'snap out of it'. One seldom has to look far to see others enduring worse conditions uncomplainingly.

It enriches one's life to develop any special aptitudes that we are fortunate to be born with. Music, handicrafts of all sorts, drawing and painting for instance. I have always wanted to draw and paint, but have never had the time or the opportunity of tuition. Now, having more time, I have taken a correspondence course in watercolour painting, from which I get a lot of fun and pleasure. It brings a new dimension into one's life, you begin to see things in colour, form and line, that you had not seen before.

Reading is one of the great pleasures of life, and it seems to me unfortunate that the doubtful pleasures of television are adversely affecting people's reading habits. We are well equipped with libraries, and 'paper backs' of all the best books are inexpensive. Here we have the work of the world's best authors, and wisdom of the ages, available to us for instruction or entertainment at a trifling cost. Do not make the mistake of undervaluing the printed word because it is cheap and easily available. It is one of the real gains to mankind from mass production.

Similarly if you have an ear for music, much pleasure can be derived from recorded music. Modern recording, and reproduction on electric players, is so good that one can enjoy the work of the world's best composers in one's own living-room at a small cost. These things to my mind are the true luxuries of life. Good health is worth some self discipline and study. I think it boils down to sensible diet regulated according to one's activities, and some daily physical exercise, preferably outdoors. However carefully diet may be regulated, I do not think one can keep physically fit without some minimum of physical work or exercise. This is too often overlooked, and there are too many 'coronarys' in middle life as a result. Moderation in all things is good. Gardening provides a lot of fun combined with mild exercise, and what better vegetables and fruit are there than those you have grown yourself?

We have all heard the term 'the pursuit of happiness' but I think this is a misnomer. Happiness can be expected to result from the right sort of living. Some of the main ingredients are good health, satisfaction in one's work, a clear conscience, pleasure in helping others, or doing public work, and the regard and affection of one's own immediate circle of family and friends. It has been truly said that 'virtue is its own reward', a fact we might emphasise a bit more in education. Not for some future prize, the good life brings a satisfaction and pleasure now, such as the inward glow one gets from some unselfish action, particularly when anonymous.

Effort, and the exercise to the full of one's endowment in brains and physique, is necessary if one is to get the most out of life and enjoy good health. It is too easy to become entrapped by the idea of the push-button existence, with its false premise that the saving of effort is necessarily a good thing.

Man's mind and body are designed for strenuous use, and they are at their best when fully used and subjected to the stresses and strains they are meant to cope with. Conversely, an organ atrophies when not used. Alexis Carrel, the famous scientist, in his book *Man the Unknown* (Hamish Hamilton) says: 'In order to reach his optimum state the human being must actualise *all* his potentialities.'

Perhaps some of what I have written may seem trite, since it has been said so often before. Never-the-less, I think this recipe for life is sound enough, based as it is on the good simple things which are so important, and I make no apology for propounding it. Western Martyr, now dead, alas, and one of my favourite authors, tried to sum it all up in one sentence—'Be gentle, brave, and cheerful, and nothing much can touch you.' I commend this to you as a sound guide to life.

Chapter 31

THE MAORIS

We who live and work amongst the Maoris see how their living habits are affected by national policy towards our native race. Matters of this kind often show end effects different from those envisaged by the politicians that initiate them in Wellington. When the Maori vote becomes of importance to the party in power, there can be undesirable results. This happened when Mr. Fraser's Labour Government was dependent on the votes of the four Maori members for its continuance in power, and was not a good thing.

The pakeha wishes the Maori well, but his efforts to give him equality and the 'fair go' that we believe in, have not always been attended by wisdom. It is not realistic to expect that a native race, existing only a hundred odd years ago as a primitive communal society, can conform in such a short period to a sophisticated western civilisation so different to their own traditional way of life. The idea that giving the Maori 'equality' is met by giving him a vote and his share of the 'handouts' in a welfare state is too easy an assumption, and one that is not helping the Maori race.

Naturally easy-going and self indulgent, the Maori has been quick to avail himself of any 'benefits' provided by the State, but seems to have failed so far to take his fair share of responsibility in community life. Is this his fault, or that of the pakeha, the 'elder brother' who should have shown more wisdom in helping the Maori to fit himself for responsibility? Money is a comparative novelty to the Maori, used as he is to a communal background, and sharing of property. Cash in hand is too often regarded as something to be spent at once. He thus tends to be somewhat improvident. He is generous, good-natured and fond of a joke, but still finds difficulty in conforming to some aspects of community life that are different to his own tribal ways.

That there is a good deal wrong in the Maoris' adjustment to the European way of living is shown by the crime rate amongst them, which is disproportionately high. Those who live amongst our native race see how much of this starts as petty crime, small thefts and suchlike, by youths who, when they leave school, have no immediate need to earn a living. Living on their families and relations, again on the communal basis, many drift into a shiftless way of life which eventually leads them into serious trouble. What seems to be lacking is some agency that could take hold of these lads when they leave school, see that they are apprenticed to some trade, or put into suitable employment with some element of discipline. Who better than the Maoris themselves to undertake this responsibility—perhaps assisted by the Mormon church, which is widely supported by the Maoris and has been such a good influence among them. Here is a chance for their leaders to show the community that they can undertake a worthwhile objective, that of curing a present state of affairs that does them no credit.

With the rapid increase in population, and the impact of large numbers of Maoris as town workers, public consciousness is being awakened to the need for more active help in fitting them to take responsibility, and adjusting themselves to a way of life very different to that of their forbears. A step forward was the recent formation of a Maori education foundation, which it is hoped will bring forward educated young Maoris to lead their own people into the better harmony between the two races that must be achieved. To the Maori that wants to make the most of himself, I think it is true to say that the average pakeha is willing to lean over backwards to help him on.

The other side of this is the Maori's response to genuine efforts to help him. The pakeha could fairly say that he would like to see a little more practical appreciation of what is being done. There are some signs of hostility towards the pakeha, evidenced in some of their truculent young men, while we would like to see their readiness to accept the considerable assistance given them matched by a bit more effort on their part. It is good to see young Maoris who have made the most of the education available to them appearing on television for instance, where they project a pleasant personality and a command of English often superior to their pakeha contemporaries. May there be many of them.

Then there is the political 'hot potato' of the Maori land question. The Treaty of Waitangi secured to the Maori the title to his land, but in the century or more that has passed since then, the question has become a tangled one. Fragmentation of ownership accelerated by the 'population explosion' has produced a situation that takes much unravelling. The practical effect of this is that there are scores, or hundreds of owners of blocks of land, the individual shares being of small value, and the land, because of multiple ownership, too often idle and useless. The Maori who has established himself as a town worker has in many cases taken on the financial responsibility of home ownership involving mortgage payments. He has no real further interest in the family land, and would be glad to sell his shares for cash to help him in his new town location. This he should be helped to do.

The dimensions of the Maori land problem are emphasised by the official figures, which show (1965) a total in both islands of 3,906,565 acres, not all of which is suitable for farming. A total of 746,000 acres of this include areas listed as of no use, suitable only for forestry, or reserves. There is general agreement that in the interests of the Maoris and the country much of this idle and unproductive land should be developed and farmed. Considerable preliminary work is involved in grouping land into suitable blocks —a recent local instance embraced sixty-seven small areas—and getting agreement of the many owners concerned. The practice of the Crown buying out the owners of fractionally small shares is part of this.

The general picture of land development today is that of taking in hand a fairly large area, and a government department, either Lands or Maori affairs, doing the work with heavy mechanical equipment. The cost is high, being given as £60 per acre for the land, with an additional £20 per acre for stock, by the Fields Director of the Department of Maori Affairs (1965).

Once developed, the enterprise represents a substantial investment of State Funds. Accepted practice today is to farm such large blocks as entities for periods of ten to twelve years for the dual purpose of recouping the cost of development, or most of it, and working the land into good heart. At this point arises the question of the future of such farms, should the land be subdivided for individual Maori farmers or should it be handed to a Trust representing the Maori owners and farmed for their benefit?

Past experience shows that, with some exceptions, the individual Maori farmer has not made a very good showing, though recent figures show that this picture is improving. Insistence that the settler qualifies himself for the job, as his pakeha counterpart has to do, and that he invests a certain minimum sum of his own as his stake in the venture, is producing better results. The alternative is to hand the farm over to a trust representative of the owners, which will farm it for their benefit. If this alternative is chosen, the Maoris should not let their pride stand in the way of making use of pakeha talent both in the management of the farm and chairmanship of such trusts. Many able and competent pakehas could help them, who have a long experience of administration and farming, and the Maori would be wise to co-opt this assistance.

As I understand it, the decision on these matters rests with the Department of Maori Affairs, and a lot depends upon the suitability or otherwise of the block for subdivision, and the availability of qualified Maoris with the prescribed minimum of capital where subdivision is contemplated. These are but the bare bones of the complex Maori land problem, which was reviewed most ably by the committee of enquiry chaired by Ivor Pritchard C.B.E. L.L.B. set up by the Minister of Maori Affairs in November 1964. This committee's recommendations, based on extensive inquiry, point the way to an accelerated solution of an unsatisfactory position.

It is to be hoped that Maori arts and crafts will be preserved, and above all that their delightful melodies and action songs will survive in their original forms, and not be corrupted by the influences of 'tourism' as has happened in Hawaii.

Our country is often cited as an example of the successful 'integration' of two races, but it would be idle to pretend that it is as yet fully successful. On the other hand, it is probably fair to say that we have made, and are still making, as honest an effort in this direction as we can.

Chapter 32

A TOUR ROUND THE FARM—ENVOI

As I come to the end of my chronicle, perhaps you who have read this far would care to accompany me for an hour or two round the farm before we part. The men are busy, and my son has asked me to move a mob of ewes from one block to another in one of the routine shifts that has to be done every few days. I am usually on call for jobs like this when required. Here is my horse Jumbo, pretending to be a bit skittish as I catch him, but really getting a bit stiff in the joints and sedate like his boss. You can ride the little mare Humbug who is a nice little thing. All saddled up and ready to go on this bright morning in early May, with a brilliant sun illuminating the lovely autumn foliage of some of the ornamental trees spotted around the place.

One more introduction and we are away—this is Spark, my half-beardie sheep dog, full of bounce and bark. A rather irrepressible kind of character who doesn't get quite enough work. As we pass the woolshed and yards you will notice some sheep yarded there. This is the 'footrot brigade', the really bad cases that have been sorted out of the flock. Their feet have been carefully pared, and they are waiting for their daily run through the foot-bath of formalin that will ultimately cure them. This is one of our chronic problems in a high rainfall area. We can cure the individual case of footrot, but what no one has yet been able to do is to cope with the continuation of hidden infection in apparently clean sheep of the flock.

This is a puzzle crying for solution, but the scientific research boys shy away from it, and so we have been unable to get any effective work done on it. We lose an estimated £2 million to £5 million a year from the effects of this disease. It affects wool growth, lambing percentages, and growth, and costs a lot to control. The scientist who could give us an answer to this would achieve world fame! Where is he?

Below the woolshed you see some new buildings. These are

quarters for shearers and shed-hands, to accommodate eight to ten men, which we built from our home-grown and milled timber last winter. They comprise three bedrooms, a spacious kitchen dining-room with electric stove and all the doings, and an ablution block with two shower rooms and a wash-house with tubs and a hot water boiler. Under New Zealand law this has to be provided for the men who shear and crutch the sheep on contract. As their period of employment seldom exceeds say three weeks in the year, it is a rather heavy item of cost in equipment of a farm such as this.

At the foot of the hill beside the road that leads to the main part of the farm is our farm sawmill, which sports a breaking down saw and a power feed breast-bench for sawing the flitches into boards and scantling. We built this ourselves from second-hand gear bought all over the place. It is roofed so that we can operate it in any weather, and is powered by our diesel-engined wheeled tractor. On the skids you will notice a number of eucalypt logs ready for cutting when it is too wet for farm work. These are drawn from a fourteen-acre plantation on the hill adjoining. We are able to keep the place supplied with timber for gates, yard rails, and maintenance of farm buildings. A lot cheaper than buying such requirements!

On our left as we continue is a two-acre plantation of pampas grass with its creamy plumes waving ten and twelve feet above the dense bushy clumps of spiky foliage not unlike our native toi-toi. The seed of this South American plant was introduced in the ballast of ships trading with South America many years ago. It has been quite widely planted as emergency forage for cattle in this northern area. Its leaves have a high fibrous content, and rather low feeding value, and cattle will not eat it unless really hungry. I planted it after losing some sixty head of cattle in a disastrous winter and spring many years ago, and it has proved its worth on one or two emergency occasions, especially the drought of 1947, since then. We are now able to save enough hay to provide for most conditions of feed shortage.

Here we come to my original abode where I lived for the first five years, rather a blot on the landscape, as it is built of corrugated iron. It now serves as a barn, and is stuffed to the roof with baled hay, which gives one a nice comfortable feeling for the approaching winter. Adjacent to it across the paddock stand our fairly extensive cattle yards. These have a dehorning bail, and are designed for drafting, with a race and gate such as we use for sheep. My son and his assistant have a big mob of cows yarded, and are engaged in the yearly culling, taking out the beasts to be sold on account of age, indifferent constitution or other defects.

Now we start to climb, using the earth roads that played their part in top dressing before the aeroplane came to our rescue. In this first small hill block are about 400 two-tooth ewes, bred on the place, running with the rams in their first breeding season. Spark gives them a whoosh together with a few exuberant barks as we move through to the next block. This is known as the Totara block, because of the predominance of young totaras in groves. These provide splendid shelter for sheep at lambing time and after shearing. Passing through this we reach the Kauri block where the ewes are, so-called because it was once dominated by the huge tree shown in the illustration facing page 112 with my horse at the foot of it. This block is a small basin of seventy acres of warm country lying to the sun which the stock like. When this area was burnt, after chopping the bush, a wet autumn resulted in a poor burn, and I well remember what a tangle of timber it was when we had to scramble through it to sow grass seed. This shows in the photograph mentioned. This has all rotted down or been burnt in heaps when it was logged up, and it is now clean pasture.

As we ride quietly to the far end I have some trouble keeping the hound quiet, but once we get behind the sheep he is full of business and bark and in short order the sheep are stringing towards the gate. It does not take long to mob them up and put them through on to a nice fresh bite of feed where their heads immediately go down as they make the most of it. This is called the Kahikatea block, because when in standing bush it had many handsome white pine trees growing in its rich damp soil.

Behind it the land rises steeply up to 1,500 feet. This is much lighter 'marginal' land which was a problem early on when we could not topdress it, and typical of large areas of bush country which have reverted in many places to scrub, fern, and second-growth. We were able to hold this in grass, and today the aircraft flying above it spreading its load of phosphate gives it a good stock carrying capacity.

Now we turn homewards, taking a different route through the Puriri block, where plenty of puriris in the standing bush indicated a fertile piece of land. For many years we worked this as a block of 150 acres, but were never able to control the steeper hills properly.

My son cut it in two with a subdivision fence, and we can now 'punch out' each section periodically with a concentration of stock. On the way, hanging round a corner in the fence, we meet a mob of Hereford calves. They are pretty little fellows with their white faces and deep red coats, but having recently been weaned they are a little disconsolate, and our appearance is greeted with a chorus of bawls, as if to say 'what have you done with mother?' They look after us with a disappointed air.

As we traverse this part, we see, ahead of us and to one side, a large area of both bush hill country and gumland, covered with scrub, all in an unimproved state, carrying nothing. Of this, some 500 acres, mostly tough broken gumland, belong to this property, and one must anticipate the question 'why hasn't it been broken in?'

The answer is, that, having developed three-quarters of the property, it is a far better bet to spend what funds we can for development on stepping up the production of existing grassland rather than start from scratch on more rough stuff. Despite tax concessions on development expenditure, so long as graduated income tax and death duties remain as high as they are, there is little attraction today in developing this land.

There it is in front of you—not less than 12,000 acres in all, in one contiguous area belonging to various owners, a monument to the stultifying effect of bad national policy. No doubt at some stage the remaining part of this property will be developed, but it looks like being a job for the grandsons.

Leaving the Puriri block we pass through a small 'junction paddock'—about an acre—where three blocks meet. We have three of these, strategically situated, and use them for holding or crossing two lots of stock over, and for mothering up ewes and lambs. The Fern paddock, which we now enter, is of a different clay soil type with a tendency to revert readily to bracken fern. When in bush it supported many big kauris, the stumps of which are still dotted around. Many of them are 5 and 6 feet in diameter, and one can imagine what majestic trees they must have been. The road running across its main face was first made by the old bushmen, and it was from this road that the loaded bullock waggon and team capsized down the hill as related in an earlier chapter.

The only effective way of dealing with bracken fern is to starve it to death by eating off its fronds and never letting it breathe. This means subdivision, constant heavy stocking and fertiliser to improve the sward.

We are nearly home as we come to the gate at the top of the first homestead paddock. To our right is the fertiliser bin, and the runway of the airstrip slopes downward and out on to a flat before us. The pilots like the downhill runway, as it enables them to get flying speed quickly with their load, and in landing it helps to pull up the empty aircraft.

Let us pause here for a few minutes while my horse breathes down my neck and Spark chases imaginary rabbits. Behind us are the hill pastures studded with sheep and cattle that were covered with rain forest when I first came here. Spread out below us are the home paddocks, a rich green from heavy topdressing, with plantations, three houses and farm buildings. This was previously desolate gum-land supporting a poor growth of scrub and fern.

As I contemplate it I have a satisfying sense of achievement and thankfulness that good health and the loyal help of others have enabled me to see it through. I reflect that I am a fortunate man—what better kind of life, giving free play to all one's instincts and abilities, could one wish for?

What we see is but a microcosm of what has taken place all over New Zealand on thousands of other farms. This is the basis of the good life available to New Zealanders. The possibilities in it are far from exhausted, and there are millions of acres either in a virgin state or capable of great improvement awaiting capital and man power. Our land has served our people well—let us see to it that this tradition is continued and that it is not starved for capital and labour in favour of less valid enterprises.

In particular let us see that farming is given its right place as a vocation for our young people, and made attractive on lines such as I have suggested. This is not being done today. My generation has done its stuff, and we must hand over to those who come after us. Let us try to see that the opportunity is given to enough of them to lead this good life to the benefit of themselves, their families, and the nation.

I have said my piece and it is now 10.30. Let us conclude this on a New Zealand note. We can lead our horses home from here—come and join me in a cup of tea.

Chapter 33

FINALE

The men and women of my generation have lived through a remarkable age in human affairs. The bicycle was a novelty about the time we were born. Heavy transport relied on steam haulage on the railways, the draught horse doing the road work. Since then we have seen the increasingly efficient internal combustion engine take over practically all forms of land transport, including the railways. The development of electricity has lightened man's labours in a thousand ways, not least in the home, and has done much to clear up the nuisance of smoke and 'smog' from the inefficient combustion of coal. Aircraft have evolved from the flimsy structures of the early days to the huge multi-engined jet air-liners of today. Our period of time witnessed the first 'moving pictures' fairly named 'the flicks' with their staccato sequences, which have become transformed into the technically superb 'talkie' in colour of today. Wireless, first as a means of communication, and then as entertainment in the home 'radio' came into use only in comparatively recent times, and is now becoming outdated by the marvels of television. The new industry of electronics comes out with a succession of surprises which will be nothing to those of the future.

I well remember my first ride in a motor car. As a boy of eight or nine I was invited to ride 'round the block' in a car, one of the first in Dunedin, owned by a lady visiting my people. It was steered by a tiller, and as I approached it with some trepidation it made a menacing hissing sound, and appeared to exude steam from every pore. I half expected it to blow up, and though thrilled by the ride, was relieved when it was over.

I had my first flight in 1917, at Moascar aerodrome in Egypt, in a Maurice Farman pusher biplane. This was a trainer machine in which the pupil sat in the nose of a flimsy plywood fuselage, with the pilot behind. The engine, directly behind the pilot, drove the wooden airscrew behind the wings. It was a clumsy-looking affair, all fabric and wires, a typical 'stringbag' of those days, but these were the machines on which pilots of World War I were trained. The flight was a real thrill, during which I took my first two photographs from the air. My pilot, Monty Turnbull, had been two years in Egypt. The following week he was shot down, with a broken leg, over a Turkish aerodrome, and taken prisoner. His wife arrived in Egypt from New Zealand that day.

It is a sad commentary on humanity that our greatly increased command over the material things of life has not been matched by a comparable advance in moral or ethical values, and that our age has seen the greatest barbarity of all times.

Although mankind in the mass does not seem to have improved very much, there are some encouraging signs. There is a universal abhorrence of war, and the United Nations, with all its imperfections, is a safeguard we have not had before. The horror of an atomic war acts as a powerful deterrent, though there is an ever present danger of the use of nuclear fission getting into the hands of some irresponsible tyrant. We have more people than ever before aware of these dangers and active in seeking to limit them.

The potentialities of atomic fission as a source of power will increasingly be developed, and should be the marvel of the age that follows ours. Let us hope that the influences working for peace may be effective against those of evil intent, in using this potential for the widespread improvement of man's lot, particularly in raising the standards of life in the under-privileged nations.

INDEX